Lecture Notes in Mathematics

1653

Editors:
A. Dold, Heidelberg
F. Takens, Groningen

Springer
Berlin
Heidelberg
New York
Barcelona
Budapest
Hong Kong
London
Milan
Paris
Santa Clara
Singapore
Tokyo

Riccardo Benedetti   Carlo Petronio

Branched Standard Spines
of 3-manifolds

Springer

Authors

Riccardo Benedetti
Carlo Petronio
Dipartimento di Matematica
Università di Pisa
Via F. Buonarroti 2
I-56127 Pisa, Italy
e-mail: benedett@dm.unipi.it
        petronio@dm.unipi.it

Cataloging-in-Publication Data applied for

Die Deutsche Bibliothek - CIP-Einheitsaufnahme

Benedetti, Riccardo:
Branched standard spines of 3 manifolds / Riccardo Benedetti ;
Carlo Petronio. - Berlin ; Heidelberg ; New York ; Barcelona ;
Budapest ; Hong Kong ; London ; Milan ; Paris ; Santa Clara ;
Singapore ; Tokyo : Springer, 1997
  (Lecture notes in mathematics ; 1653)
  ISBN 3-540-62627-1
NE: Petronio, Carlo:; GT

Mathematics Subject Classification (1991): Primary: 57N10
                                            Secondary: 57M20, 57R15, 57R25

ISSN 0075-8434
ISBN 3-540-62627-1 Springer-Verlag Berlin Heidelberg New York

Typesetting: Camera-ready T$_E$X output by the authors
SPIN: 10520303   46/3142-543210 - Printed on acid-free paper

Acknowledgements

This research was carried out while the first named author was a visiting professor at the Institut Fourier of Grenoble and at the Université Louis Pasteur of Strasbourg, and he would like to thank these institutions for their hospitality. The second named author thanks for travel support the following institutions: the Centre Emile Borel of Paris, the University of Oporto and the University of Bochum. Both authors acknowledge partial support by the Italian MURST 40% research project "Geometria reale e complessa"

We warmly thank Vladimir Turaev, Joe Christy, Alexis Marin and Oleg Viro for helpful conversations or for having shown interest in our work.

Contents

1 Motivations, plan and statements — 1

1.1 Combinatorial realizations of topological categories 1

1.2 Branched standard spines and an outline of the construction 3

1.3 Graphic encoding 5

1.4 Statements of representation theorems 5

1.5 Existing literature and outline of contents 10

2 A review on standard spines and o-graphs — 13

2.1 Encoding 3-manifolds by o-graphs 13

2.2 Reconstruction of the boundary 17

2.3 Surgery presentation of a mirrored manifold and ideal triangulations . . 20

3 Branched standard spines — 23

3.1 Branchings on standard spines 23

3.2 Normal o-graphs 26

3.3 Bicoloration of the boundary 28

3.4 Examples and existence results 32

3.5 Matveev-Piergallini move on branched spines 37

4 Manifolds with boundary — 40

4.1 Oriented branchings and flows 40

4.2 Extending the flow to a closed manifold 45

4.3 Flow-preserving calculus: definitions and statements 47

4.4 Branched simple spines 50

4.5 Restoring the standard setting 55

4.6 The MP-move which changes the flow 60

5 Combed closed manifolds — 64

5.1 Simple vs. standard branched spines 64

5.2 The combed calculus 69

6 More on combings, and the closed calculus — 73

6.1 Comparison of vector fields up to homotopy 73

6.2 Pontrjagin moves for vector fields, and complete classification 76

6.3 Combinatorial realization of closed manifolds 81

7 Framed and spin manifolds   85
  7.1   The Euler cochain   85
  7.2   Framings of closed manifolds   87
  7.3   The framing calculus   91
  7.4   Spin structures on closed manifolds   94
  7.5   The spin calculus   95

8 Branched spines and quantum invariants   98
  8.1   More on spin structures   98
  8.2   A review of recoupling theory and Reshetikhin-Turaev-Witten invariants   99
  8.3   Turaev-Viro invariants   101
  8.4   An alternative computation of TV invariants   104

9 Problems and perspectives   108
  9.1   Internal questions   108
  9.2   Questions on invariants   110
  9.3   Questions on geometric structures   116

10 Homology and cohomology computations   121
  10.1 Homology, cohomology and duality   121
  10.2 More homological invariants   123
  10.3 Evenly framed knots in a spin manifold   125

Bibliography   127

Index   131

Chapter 1

Motivations, plan and statements

In this chapter we will describe the general plan of our work and state in full detail our main results. However the reader should refer to the subsequent chapters for a correct interpretation of the *effectiveness* of the maps whose existence is stated in Theorems 1.4.1, 1.4.2, 1.4.3 and 1.4.4. It is in the reconstruction process which underlies the definition of these maps that branched standard spines come into play.

1.1 Combinatorial realizations of topological categories

The main point of this work is the construction of combinatorial realizations of some categories of 3-manifolds with extra structure. Before describing in detail the classes of objects which we handle, let us explain what we exactly mean by *combinatorial realization*. Let $\mathcal{T}$ be a collection of topological objects (for us, 3-dimensional manifolds with a certain extra structure), regarded up to a suitable equivalence relation. The first ingredient of a realization will be an explicitly described set $\mathcal{S}$ of finite combinatorial objects (typically, finite graphs with certain properties and decorations), together with an effective mapping

$$\Psi : \mathcal{S} \to \mathcal{T},$$

called the *reconstruction map*, whose image is the whole of $\mathcal{T}$. The second ingredient of the realization is the *calculus*, namely an explicitly described finite set of local moves on $\mathcal{S}$ with the property that two elements of $\mathcal{S}$ have the same image in $\mathcal{T}$ under Ψ if and only if they are related to each other by a finite combination of the moves of the calculus.

A classical model of a combinatorial realization is the presentation of links up to isotopy in S^3 by means of planar diagrams, where the calculus is generated by Reidemeister moves. In 3-manifold topology some examples of combinatorial representation are known which satisfy at least some of the requirements which we have stated. Let us remark however that the requirements of *finiteness* and *locality* of the calculus are somewhat restrictive. For instance the presentation of closed connected oriented 3-manifolds via longitudinal Dehn surgery on framed links in S^3, with either of the two versions of the Kirby calculus, does not satisfy the requirements. If we use the version of the calculus which includes the band move, then we have a non-local move, whereas, if we use the generalized Kirby move, then we have indeed local moves, but we actually have to take into account infinitely many different ones, parametrized by the number of strands which link the curl removed by the move. In [6], using a slight refinement of the theory of standard spines and an appropriate graphic encoding, we have produced a combinatorial realization of the category of compact connected 3-manifolds with non-empty boundary, and a refined version of the same realization for the oriented case. This realization will be recalled in Chapter 2 because we will use it extensively.

Let us now describe the topological objects of which we provide a combinatorial realization in this work. By a 3-manifold we will always mean a *connected, compact* and *oriented* one, with or without boundary. We will denote by $\mathcal{M}$ the class of *closed* 3-manifolds, up to orientation-preserving diffeomorphism. We will call *combing* on a closed 3-manifold a nowhere-vanishing vector field, and *framing* a triple of linearly independent vector fields which pointwise induce the orientation. Combings and framings will always be viewed up to homotopy through objects of the same type. Other objects which we will consider are spin structures. Note that if $f : M \to M'$ is an orientation-preserving diffeomorphism then to any combing or framing or spin structure on M there corresponds under f an object of the same type on M'. We will denote by $\mathcal{M}_{\text{comb}}$ the set of pairs (M, v) where M is a closed manifold and v is a combing on M, viewed up to the natural action on pairs of orientation-preserving diffeomorphisms. We define $\mathcal{M}_{\text{fram}}$ and $\mathcal{M}_{\text{spin}}$ in a similar way. For each of $\mathcal{M}$, $\mathcal{M}_{\text{comb}}$, $\mathcal{M}_{\text{fram}}$ and $\mathcal{M}_{\text{spin}}$ we will give a combinatorial realization as stated above (the realization of $\mathcal{M}$ being different and independent of the one given in [6]).

Before sketching the topological constructions which underlie our presentations, we want to point out a reason for which, to our opinion, it is non-trivial and of some interest to give a combinatorial realization of the classes of "structured manifolds" $\mathcal{M}_{\text{comb}}$, $\mathcal{M}_{\text{fram}}$ and $\mathcal{M}_{\text{spin}}$. Our remark is that, for any given closed manifold M, the set of extra structures of each of the three types in exam can be described as an *affine space* (or a certain "bundle" of affine spaces, in the case of combings), and there is no canonical way to define a preferred basepoint. This is well-known for spin structures, which are identified to an affine space over $H^1(M; \mathbb{Z}_2)$. Once a preferred framing on M is fixed, the set of all framings can be identified to the set of homotopy classes of maps from M to $SO(3)$, which is rather easily recognized to be a central extension of $H^1(M; \mathbb{Z}_2)$ by $\mathbb{Z}$; however the dependence on the choice of the basepoint is somewhat intriguing. The situation is more complicated for the case of combings: if we fix a framing on M then the set of all combings can be identified to the set of homotopy classes of maps $v : M \to S^2$; if h denotes the preferred generator of $H^2(S^2; \mathbb{Z})$ we can associate to v the element $g(v) = v^*(h) \in H^2(M; \mathbb{Z})$, and the map g is surjective. Moreover, for $k \in H^2(M; \mathbb{Z})$ and $v, v' \in g^{-1}(k)$, the difference between v and v' can be described as a "Hopf number", which lies in $\mathbb{Z}$ when k is a torsion element, and in $\mathbb{Z}_{2d}$ when k is not a torsion element and d is the greatest integer divisor of k. So, if we fix a basepoint in each of the fibres of g, then we can identify the fibre with the appropriate $\mathbb{Z}$ or $\mathbb{Z}_{2d}$, but the choice of these basepoints, just as the choice of the framing from which we have started, cannot be made canonical.

To our point of view, heuristically speaking, it follows from the facts remarked in the previous paragraph that combinatorial realizations of $\mathcal{M}_{\text{comb}}$ or $\mathcal{M}_{\text{fram}}$ or $\mathcal{M}_{\text{spin}}$ must be supported by classes of objects capable of capturing deep topological properties of 3-manifolds. We have devised such a class of objects by introducing *branched standard spines*. This notion has been obtained as a combination of the two essentially classical concepts of *standard spine* and of *branched surface*, and this makes our definition a rather natural one.

The interest into effective combinatorial presentations was increased in recent years by the development of the theory of quantum invariants (see e.g. [56]). On one hand the existence and structure of these invariants has been predicted, starting from Witten's interpretation of the Jones polynomial, from general principles in quantum field theory [61]; but on the other hand an effective and rigorous construction of the invariants has only been given, according to the current opinion, via combinatorial presentations such as surgery with the Kirby calculus (for the Reshetikhin-Turaev-Witten invariants), or spines and triangulations with the appropriate moves (for the Turaev-Viro invariants). Our work was partially influenced and motivated by this consideration, and we show in Chapter 8 that our combinatorial presentation of spin manifolds is suitable for an effective implementation and computation of the spin-refined version of the Turaev-Viro invariants.

In the last few years new models for invariants have been proposed (see e.g. [23], [29], [33]) which are based on ideas similar to those used for the quantum invariants (i.e. representations of certain classes of diagrams into algebraic objects, typically Hopf algebras with extra structures), but do not strictly fall in the quantum category (as axiomatized for instance in [56]). The invariants constructed by G. Kuperberg [33] are in their most general form invariants of framed and combed manifolds, and moreover their definition does not immediately yield a computation recipe; therefore it is very natural ask if our combinatorial presentations support an effective implementation of the invariants. We discuss this question in greater detail in Chapter 9, where we also mention other fields in which our work might have applications. In particular, after remarking that our combed calculus dually represents the set of homotopy classes of oriented plane distributions on 3-manifolds, we give some hints on the possible relations with the theory of foliations and contact structures (recall that every homotopy class of oriented plane fields contains both distributions and contact structures).

1.2 Branched standard spines and an outline of the construction

The authors who mainly contributed to developing the theory of standard spines are Casler [13], Matveev [37] and Piergallini [45]; we also remind the reader that on this theory is based the combinatorial realization of manifolds with boundary given in [6]. We warn the reader that our definition of *standard* spine includes the requirement that the strata of the natural stratification should be cells; this cellularity condition is in our opinion essential to base a *local* calculus on spines, and we will point out its importance in various instances. We will use the term *simple* for spines in which the cellularity condition is dropped.

The notion of branched surface was originally introduced in [60] for the study of hyperbolic attractors (see also [11]). Later on, in a number of papers (e.g. [17], [19] and the references quoted therein) branched surfaces have been viewed as codimension 1 objects capable of supporting and generalizing classical notions about surfaces, in particular incompressibility.

A branched standard spine P of a (compact, connected and oriented) 3-manifold N with non-empty boundary is a standard spine of N endowed with an oriented branching, which roughly speaking means that pointwise an oriented tangent plane to P is well-defined, and the local behaviour can be described by certain very natural models. To every such a P we can associate a vector field on N (well-defined up to a certain type of homotopy) which is positively transversal to P and has a prescribed local behaviour on a neighbourhood of ∂N; namely, the field is tangent to ∂N only along finitely many simple curves of *apparent contour*, along which the boundary is "concave" with respect to the flow. This theory specializes to the closed case by requiring that N should be bounded by S^2 and the field should have the simplest possible behaviour near the boundary. This allows to extend the field to the closed manifold M obtained by capping off the boundary sphere of N, and the field on M turns out to be well-defined up to homotopy. Since we can prove that all combings on M are obtained via this procedure from some P, we have the reconstruction map. The moves of the calculus for combings are essentially branched versions of the usual moves of the Matveev-Piergallini calculus for standard spines. However we want to emphasize that the results known for the non-branched case do not imply the fact that the combing moves generate the equivalence relation induced by the reconstruction map, and our proof that this indeed happens is based on an independent argument.

As we already mentioned, the natural realm in which the foundations of the theory of branched standard spines are placed is the category of 3-manifolds with boundary, of which the closed case is a specialization. The same is true case also for the core of the combing calculus, which is first established in the case with boundary and then specialized. Moreover the proof is divided into two steps; at first we only use general position arguments to derive a weaker version of the calculus, in which spines may not be standard and therefore moves are non-local; later a harder work allows us to recover the standard context and hence locality.

In the closed case we show how to complete the calculus for combings by means of a local move, which we call *combinatorial Pontrjagin move*, whose topological counterpart allows to obtain from each other any two combings on the same manifold. This leads therefore to a new combinatorial realization of (connected, oriented) closed 3-manifolds.

The combinatorial realization of combings allows us to deduce a similar one for framings, starting with the following (easy) remarks: the first vector of a framing on a closed manifold M is a combing, and a combing v on M extends to a framing if and only if the Euler class $\mathcal{E}(v) \in H^2(M; \mathbb{Z})$ of the plane field complementary to v is null. Our next step is to construct for a branched standard spine P of M a preferred cocycle c_P which represents the Euler class of the combing carried by P. Then we show how to explicitly associate to an integral 1-cochain whose coboundary is c_P a well-defined framing, and we prove that two such cochains define the same framing if and only if the reduction modulo 2 of their difference is a $\mathbb{Z}_2$-coboundary. Therefore the objects of the combinatorial realization of framed manifolds are pairs (P, x), where x is a mod-2 1-cochain on P which lifts to an integral 1-cochain $\tilde{x}$ with $\delta\tilde{x} = c_P$, and x is viewed up to $\mathbb{Z}_2$-coboundaries. The moves of the framing calculus are obtained by enhancing to such pairs the moves of the combing calculus, making sure that whenever (P, x) and (P', x') are related by a move then they define the same framing.

A similar scheme works for spin structures. Such a structure on a manifold M can be viewed as a framing of M on the singular set $S(P)$ of a spine P of M, where the framing extends to the whole of P and is viewed up to homotopy on $S(P)$. Given a certain branched spine P and the corresponding combing v, the second Stiefel-Whitney class of the plane field complementary to v is just the reduction modulo 2 of $\mathcal{E}(v)$, therefore it coincides with $w_2(M)$ and so it is null; moreover $H^1(P; \mathbb{Z}_2) = H^1(M; \mathbb{Z}_2)$, and it easily follows that every spin structure has a representative whose first vector is v. To every mod-2 1-cochain x on P whose coboundary is the reduction modulo 2 of c_P we can associate a spin structure which depends only on the class of x modulo $\mathbb{Z}_2$-coboundaries. Therefore spin 3-manifolds are encoded by such pairs (P, x), and the moves to take into account are the same as in the case of framings (which is coherent with the fact that each framing determines a spin structure), together with suitable enhancements of the combinatorial Pontrjagin move of the closed calculus.

1.3 Graphic encoding

Another element of our work is a graphic encoding of the objects of the representations. This is conceptually subordinate to the intrinsic theory of branched spines, but not secondary if one has in mind the effectiveness (and possibly an actual computer implementation) of the combinatorial realizations. To a branched standard spine we associate a planar graph by embedding in 3-space a neighbourhood of the singular set of the spine, in such a way that along the singular set the tangent plane is "almost horizontal", and then projecting on the horizontal plane. The (oriented) branching allows to define this operation uniquely, and implies that to every graph with some simple extra structures there corresponds a unique branched standard spine, and moreover the decoding procedure is completely explicit and effective.

From a given graph it is very easy to explicitly reconstruct the 2-cells of the corresponding spine, and also to find the preferred cochain which represents the Euler class of the combing carried by the spine. Moreover the graph itself corresponds to the singular set of the spine, so a 1-cochain on the spine can be viewed just as a colouring of the edges of the graph. These facts imply that the various elements which interact in the combinatorial realization of spin and framed 3-manifolds can be very easily translated in terms of the graphs.

We also want to note that on one hand the graphs still carry in a rather transparent way all the information on the topological situation, and on the other hand they allow a formal control of the various manipulations (see also [7]). In Chapter 10 we will show how to explicitly carry out the computation of some homological invariants using the graphs only.

1.4 Statements of representation theorems

In this section we will describe in a completely self-contained way, i.e. without referring to the notions described in the previous sections, our *combinatorial realizations* of $\mathcal{M}$, $\mathcal{M}_{\text{comb}}$, $\mathcal{M}_{\text{fram}}$ and $\mathcal{M}_{\text{spin}}$. As already pointed out, the reader should refer to the rest of this work to understand correctly the *reconstruction* as a two-step process, which takes from a graph to a branched standard spine and then to a 3-manifold with extra structure.

We start by defining the combinatorial objects of our realizations. Before doing this we adopt a useful convention for the description of graphs. We will be dealing in the sequel with abstract finite graphs in which every vertex has a neighbourhood with

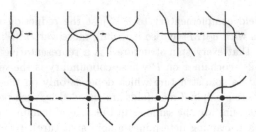

Figure 1.1: Reidemeister-type moves

a prescribed embedding in the plane. Our convention is that *we will always describe the whole graph by a generic immersion in the plane, marking by solid dots the genuine vertices, to distinguish them from the double points of the immersion.* Equivalently, we will refer to planar graphs and view them up to planar isotopy and the Reidemeister-type moves of Figure 1.1, which express the fact that the vertices which are not marked by solid dots are actually fake vertices.

Axioms for normal o-graphs. We will denote by $\mathcal{N}$ the set of finite connected quadrivalent graphs Γ with the following properties:

N1. In Γ there is at least one vertex, and every vertex has a neighbourhood which is embedded in the plane as a normal crossing, where the embedding is viewed up to planar isotopy. Moreover two opposite germs of edges incident to the vertex are marked as being over the other two, as in link projections;

N2. Each edge has a direction, and the directions of opposite edges match through the vertices.

Note that, if one chooses to view o-graphs as genuinely planar graphs with fake crossings, then one has to be careful and remember that non-marked crossings do not break edges.

The set of normal o-graphs will be used to represent manifolds with boundary, and the following subset will correspond to closed manifolds:

Axioms for closed normal o-graphs. We will denote by $\mathcal{G}$ the set of elements Γ of $\mathcal{N}$ which satisfy also:

C1. If one removes the vertices and joins the edges which are opposite to each other, the result is a unique (oriented) circuit;

C2. The trivalent graph obtained from Γ by the rules of Figure 1.2 is connected.

C3. Consider the disjoint union of oriented circuits obtained from Γ by the rules of Figure 1.3. Then the number of these circuits is exactly one more than the number of vertices of Γ.

Since the same objects will be used to encode combed closed manifolds and closed manifolds without extra structures, we set $\mathcal{G}_{\text{comb}} = \mathcal{G}$. Before proceeding we note that, given an o-graph Γ and one of the oriented circuits γ of Figure 1.3, we can associate

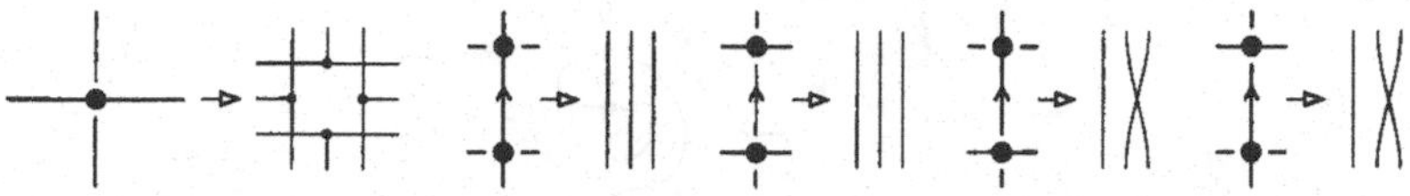

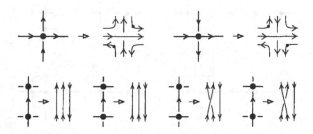

Figure 1.2: A trivalent graph associated to a normal o-graph (recall that non-marked vertices are fake)

Figure 1.3: A union of circuits associated to a normal o-graph. The orientations and the solid dots on the circuits in this figure are relevant only for the definition of spin and framed normal o-graphs

in a natural way to γ an oriented loop $e_1^{\alpha_1} \cdots e_p^{\alpha_p}$ in Γ, where $e_1, \ldots, e_p$ are edges of Γ (possibly with repetitions) and $\alpha_1, \ldots, \alpha_p \in \{\pm 1\}$ depend on the matching of orientations. We also note that on each circuit there is necessarily an even number of solid dots (there is an intrinsic reason for this, but a combinatorial proof is also easy).

Axioms for framed normal o-graphs. We will denote by $\mathcal{G}_{\mathrm{fram}}$ the set of graphs Γ which satisfy axioms N1, N2, C1, C2, C3 and the following ones:

F1. To each edge of Γ is attached an element of $\mathbb{Z}_2$, called the colour of the edge;

F2. It is possible to attach an integer to each edge so that:

    (a) On each edge, the class of this integer modulo 2 is the colour of the edge;

    (b) The following should happen for each of the circuits γ of Figure 1.3: if the oriented loop $e_1^{\alpha_1} \cdots e_p^{\alpha_p}$ corresponds in Γ to γ and the integer x_i is attached to the edge e_i, then on γ there should be $2(1 - \sum_{i=1}^p \alpha_i x_i)$ solid dots.

Axioms for spin normal o-graphs. We will denote by $\mathcal{G}_{\mathrm{spin}}$ the set of graphs Γ which satisfy axioms N1, N2, C1, C2, C3 and the following ones:

S1. To each edge of Γ is attached an element of $\mathbb{Z}_2$, called the colour of the edge;

S2. The following should happen for each of the circuits γ with solid dots of Figure 1.3: let $e_1^{\alpha_1} \cdots e_p^{\alpha_p}$ be the loop in Γ which corresponds to γ, let the element x_i of $\mathbb{Z}_2$ be attached to the edge e_i and assume that on γ there are $2n$ solid dots. Then $\sum_{i=1}^p x_i \equiv n + 1 \mod 2$.

Figure 1.4: First move of the combing calculus

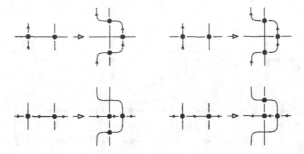

Figure 1.5: More moves of the combing calculus

We explain now some conventions which we use in describing local moves to be performed on our combinatorial objects. We always show only a portion of graph, with the agreement that the graph should remain unaltered outside the region shown. If the orientation of some portion of graph is not shown, then the portion can be oriented arbitrarily and should be oriented accordingly after the move (therefore, several different combinatorial moves can be encoded by the same figure). In the case of $\mathcal{G}_{\text{spin}}$ and $\mathcal{G}_{\text{fram}}$, as a result of the application of some of the moves, some edge might have multiple colours in $\mathbb{Z}_2$, and the convention is that they should be summed up in $\mathbb{Z}_2$.

Theorem 1.4.1. *There exists an effective surjective reconstruction map*

$$\Phi_{\text{comb}} : \mathcal{G}_{\text{comb}} \to \mathcal{M}_{\text{comb}},$$

and the equivalence relation defined by Φ_{comb} is generated by the local moves of Figures 1.4 and 1.5. Moreover all these moves automatically preserve all the axioms of the objects $\mathcal{G}_{\text{comb}}$.

Theorem 1.4.2. *There exists an effective surjective reconstruction map*

$$\Phi : \mathcal{G} \to \mathcal{M},$$

and the equivalence relation defined by Φ is generated by the move of Figure 1.6 together with the same moves of Figures 1.4 and 1.5 as in the combed case. Again, the moves automatically preserve the axioms of $\mathcal{G}$.

Theorem 1.4.3. *There exists an effective surjective reconstruction map*

$$\Phi_{\text{fram}} : \mathcal{G}_{\text{fram}} \to \mathcal{M}_{\text{fram}},$$

and the equivalence relation defined by Φ_{fram} is generated by the local moves of Figures 1.7 and 1.8. These moves automatically preserve the axioms of $\mathcal{G}_{\text{fram}}$.

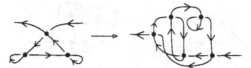

Figure 1.6: Last move of the closed calculus

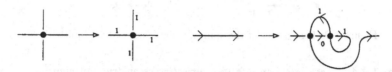

Figure 1.7: First move of the framing and spin calculus

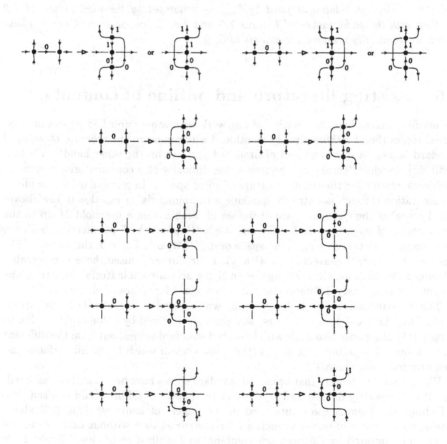

Figure 1.8: More moves of the framing and spin calculus

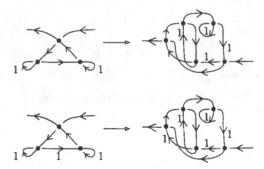

Figure 1.9: Last moves of the spin calculus (colours 0 are omitted)

Theorem 1.4.4. *There exists an effective surjective reconstruction map*

$$\Phi_{\text{spin}} : \mathcal{G}_{\text{spin}} \to \mathcal{M}_{\text{spin}},$$

and the equivalence relation defined by Φ_{spin} is generated by the local moves of 1.9 together with the same moves of Figures 1.7 and 1.8, as in the framed case. These moves automatically preserve the axioms of $\mathcal{G}_{\text{spin}}$.

1.5   Existing literature and outline of contents

We briefly describe here the relation of our work with some published papers on connected topics (besides those already mentioned which concern the separate theories of standard spines on one hand and of branched surfaces on the other hand). We first recall [55], in which Turaev first addresses the difficulty of a combinatorial treatment of objects which have the intrinsic nature of affine spaces. In particular he provides a representation (though not strictly speaking a combinatorial realization in the above-stated sense) of the quotient space of the set of combings on a manifold M up to the identification of combings which differ only for a "Hopf number" (as mentioned above, the outcome is in this case an affine space over $H^2(M; \mathbb{Z})$). Some of the ideas of [55], together with several conversations with Vladimir Turaev himself, have considerably influenced our treatment of framings (even if the actual combinatorial objects of the calculus are completely different, since [55] does not refer to branched spines).

The construction of the Euler cochain, which plays a central role in the treatment of framings and spin structures, was partially inspired by a construction due to Christy [11]; this paper also deals with branched standard spines, but from the different point of view of hyperbolic attractors (the framework in which branched surfaces had been originally introduced).

We also want to remark that branched standard spines have been tacitly considered, even if not explicitly defined and studied, in the papers of Gillman and Rolfsen [20], [21] about the Zeeman conjecture, and in the papers of Ishii [24], [25], [26] about "flow-spines". Some of our constructions for the closed case without extra structure are partially inspired by [26], and our combinatorial realization of closed 3-manifolds (Theorem 1.4.2) is actually a solution *in a standard setting* of the same problem faced in [26]. Several considerations spread over the text (and concentrated in Section 5.1) have suggested that a different approach to this problem might be advisable. Note in particular that the graphs by which we represent spines are *not* those of [26]; a remarkable difference is that our graphs are completely described by elementary combinatorial conditions, while in [26] it is declared that not all the graphs under consideration actually represent spines, and this extra condition is not translated in combinatorial terms.

We note that for our aims it was natural to consider transversely oriented branched spines of oriented manifolds, but we provide the main foundational definitions also for the non-oriented case. One could try to extend the whole theory to this more general setting, and it is conceivable that some of the constructions of this work can be extended or adapted. We will confine just to a few remarks and examples.

Let us now describe more precisely the topics treated in the individual chapters. In Chapter 2 we review the theory of standard spines of 3-manifolds with boundary, including the Matveev-Piergallini calculus, and introduce, following [6], a combinatorial encoding of standard spines (and, therefore, manifolds) by means of objects which we call o-graphs; we consider here only the oriented case, addressing the reader to [6] for the more complicated general case. Next, we show how to determine the boundary directly from an o-graph. Moreover, following J. Roberts ([47], see also [7]) we explain how to construct, starting from an o-graph of a manifold N, a framed link in S^3 which represents $N \sqcup_\partial (-N)$ by surgery.

In Chapter 3 we introduce the combinatorial notion of branching on a standard spine, providing some motivations and interpretations. After restricting our attention to oriented branchings, we show that a spine with such a structure admits a canonical o-graph; moreover we discuss several examples, we prove that every 3-manifold with boundary has spines which can be given an oriented branching, and we show that a branched standard spine induces a certain splitting (called bicoloration) of the boundary of the corresponding manifold. In the last section we analyze, from a purely combinatorial point of view, the effect on branchings of the fundamental move of the Matveev-Piergallini calculus.

In Chapter 4 we develop a geometric interpretation of the notion of oriented branching, showing that on a 3-manifold with boundary a branched standard spine corresponds to a flow with certain well-understood properties (in particular, every orbit should start and end on the boundary). This allows, among other things, a natural interpretation of the bicoloration of the boundary. In the second section we prove various elementary facts concerning the extension of a flow from a manifold with boundary to a closed manifold. Next, we show that two flows carried by branched spines are homotopic through flows of the same qualitative type if and only if the corresponding spines are obtained from each other by means of certain combinatorial moves. This result is obtained by first relaxing the notion of standard spine (by dropping the cellularity condition), and then restoring the standard setting by an accurate analysis of the moves. In the last section we show that a certain additional move allows to obtain branched standard spines representing all homotopy classes of flows, but we explain why this is not sufficient to deduce a new calculus for 3-manifolds with boundary.

In Chapter 5 we start analyzing the case of closed manifolds, which emerges by considering spines of manifolds bounded by S^2, with branchings inducing the splitting of S^2 into two discs. After recalling the notion of flow-spine due to I. Ishii, we make various comments on its relation with the notion of oriented branching on a standard spine. In particular we show that there exist non-isomorphic flow-spines with isomorphic neighbourhood of the singular set, which implies that the singularity data are not sufficient to encode the spine. Then we proceed to the proof of Theorem 1.4.1, which provides the combinatorial realization of combed manifolds. The proof combines some of Ishii's ideas with the technical machinery developed in the previous chapter.

Chapter 6 is devoted to the combinatorial realization of closed manifolds based on branched standard spines (Theorem 1.4.2). This requires a detailed analysis of the set of homotopy classes of non-singular vector fields on a 3-manifold (in particular, we sketch a complete classification). As a result of this analysis we provide a transformation rule for vector fields (called Pontrjagin move) iterating which one can obtain from each other any two homotopy classes of vector fields. In the last section we complete the proof of Theorem 1.4.2 by translating the Pontrjagin move into a (rather complicated) combinatorial move on o-graphs.

Chapter 7 concerns the calculi for framed and spin manifolds (Theorems 1.4.3 and 1.4.4). In the first section we show that there exists a preferred cochain representing the Euler class of the plane distribution complementary to the flow carried by a branched spine. The combinatorial realization of framed manifolds is then obtained by restricting to branched spines with null Euler class and adding extra data which allow to encode the second vector of the framing. This requires a detailed analysis of the set of framings up to homotopy on a given 3-manifold, of which we also provide a complete classification; the technically more complicated step of the proof of Theorem 1.4.3 consists in enhancing the moves of the combing calculus to framed moves. For Theorem 1.4.4 we proceed in a similar way, showing that certain extra data on branched spines allow to encode a spin structure, and translating the moves of the closed calculus into spin moves.

In Chapter 8 we recall the definitions of spin-refined quantum invariants (Reshetikhin-Turaev-Witten and Turaev-Viro) and show that the environment of our combinatorial realization of spin manifolds allows an effective description of these invariants and of the various versions of the Turaev-Walker theorem. This chapter contains essentially no new ideas, and follows quite closely the approach of J. Roberts [47], [48] to Turaev-Viro invariants.

In Chapter 9 we list and motivate several problems which arise in connection with the theory of branched standard spines. After mentioning some questions which strictly concern our combinatorial realizations, we turn to invariants; in this context we devote some space to a recent construction due to G. Kupenberg [33] of invariants of combed and framed 3-manifolds, outlining the possible relevance of branched standard spines to an effective implementation of this construction. We close the chapter with some questions concerning the relation of branched standard spines with foliations and contact structures on 3-manifolds.

Chapter 10 is essentially an appendix to the main theme of this monograph; it contains some homology and cohomology computations (which are used in various instances in the book), carried out by means of branched standard spines. Moreover we redefine and analyze some other homological invariants, in particular for the case of spin manifolds.

Chapter 2

A review on standard spines and o-graphs

In this chapter we recall the main results and constructions of [6], [43] and [7]. Namely we review the notion of (oriented) standard polyhedron and its relation with the theory of (oriented) 3-manifolds, we introduce the graphic encoding for oriented standard polyhedra by means of o-graphs, and we quickly describe the correspondence between o-graphs and manifolds, including the computation of the boundary.

2.1 Encoding 3-manifolds by o-graphs

When dealing with polyhedra and topological manifolds we will work in the PL category (and in this context, without mention, view objects up to PL-isomorphism). Later in this work we will need to switch to a smooth viewpoint (Section 4.1).

A finite 2-dimensional polyhedron is called *simple* if every point in it has a neighbourhood homeomorphic to an open subset of

$$(\mathbb{R}^2 \times \{0\}) \cup (\mathbb{R} \times \{0\} \times \mathbb{R}_+) \cup (\{0\} \times \mathbb{R} \times \mathbb{R}_-) \subset \mathbb{R}^3.$$

The typical local shapes are those shown in Figure 2.1.

A simple polyhedron P is naturally stratified as $V(P) \subset S(P) \subset P$, where $V(P)$ is a finite set of points called vertices (the points which play the role of 0 in the above subset of $\mathbb{R}^3$), and $S(P)$ is the singular set (the set of points which play the role of $(t,0,0)$ or $(0,t,0)$). The simple polyhedron P is called *standard* if the components of $S(P) \setminus V(P)$ are (open) segments (called edges), and those of $P \setminus S(P)$ are (open) discs.

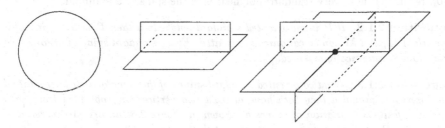

Figure 2.1: Local shapes in a simple polyhedron

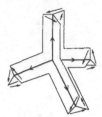

Figure 2.2: A compatible orientation near a vertex

The reason why standard polyhedra are interesting is that based on them there exists a calculus for 3-manifolds with boundary [13], [37], [45]. This calculus has been slightly refined to oriented 3-manifolds in [6], and translated into a planar graphic calculus. Let us very briefly sketch how this calculus goes. A standard polyhedron P is called a *spine* of a 3-manifold with boundary M if P embeds in the interior of M so that M collapses onto P. For every M such P's exist and determine M uniquely [13], and two P's of the same M are related by certain known moves [37], [45]. Not every P is the spine of some M, but every *oriented* P (as defined in [6] and recalled below) is the spine of some oriented M, and every P of an oriented M is naturally oriented. Moreover oriented P's have an easy graphic description in terms of decorated planar graphs. In the rest of this section we will expound this theory in greater detail.

In a standard polyhedron, we will speak of *germ* of edge at a vertex, germ of disc at a vertex, and germ of disc at an edge, in an obvious intuitive sense which we leave to the reader to formalize. Every edge defines 2 germs (possibly at the same vertex) and at every vertex there are 4 germs of edge. At every vertex there are 6 germs of disc, at every edge there are 3 germs of disc, whereas a disc might define any number of germs.

Definition 2.1.1. An *orientation* for a standard polyhedron is the choice of a direction for each edge and a cyclic order for the 3 germs of disc at it, well-defined up to simultaneous reversal of the edge-direction and the disc-ordering, with the compatibility condition described in Figure 2.2.

This definition is motivated by the following results, for which we address the reader to [6]; recall that not every standard polyhedron is the spine of a manifold.

Proposition 2.1.2. *If P is an oriented standard polyhedron then P is the spine of a 3-manifold M, and M can be canonically oriented. Every standard spine of an oriented 3-manifold is canonically oriented.*

Proposition 2.1.3. *Any two oriented standard spines of the same oriented 3-manifold with boundary, provided they both have at least two vertices, are obtained from each other by a finite combination of moves as shown in Figure 2.3 (an oriented version of the* Matveev-Piergallini *move) and inverses of this move.*

We now define the objects by which we represent manifolds, state the representation theorem and outline the construction which leads to it. When dealing with graphs we will use the same conventions as stated in Chapter 1, namely we will use planar immersions and mark the genuine vertices by solid dots.

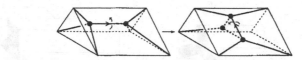

Figure 2.3: The oriented Matveev-Piergallini move

Figure 2.4: These moves are called respectively C and MP

Definition 2.1.4. An *o-graph* is a connected finite quadrivalent graph Γ with the following properties and extra structures:

1. There is at least one vertex;

2. Every vertex has a neighbourhood with an assigned embedding in the plane as a normal crossing, and two opposite branches at the crossing are marked as being over the other two, as in link projections;

3. An element of $\mathbb{Z}_3$ (called the colour) is attached to every edge of Γ.

We will always refer to Γ itself as an o-graph.

Recall from Section 1.4 that we always use planar graphs to represent abstract ones, keeping in mind that crossings which are not marked are not really part of the structure. Alternatively one could use planar graphs viewed up to the moves of Figure 1.1.

Theorem 2.1.5. *Compact, connected, oriented 3-manifolds with non-empty boundary correspond bijectively to o-graphs with at least two vertices up to the equivalence relation generated by the moves shown in Figure 2.4.*

In Figure 2.4 the convention is that an edge can be equivalently given a colour α or several colours (each localized at some subsegment) whose sum is α. (See below Figure 3.10 for explicit examples of how to apply the moves.) The following are the constructions which lead to the theorem.

FROM AN O-GRAPH TO A (UNIQUE) 3-MANIFOLD. Fix Γ. To every marked vertex associate a 2-polyhedron as shown in Figure 2.5 (left). This polyhedron terminates

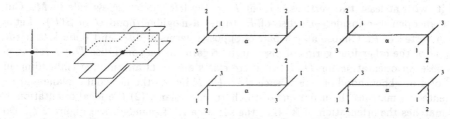

Figure 2.5: The polyhedron introduced for a vertex and the numbering the legs of the T's

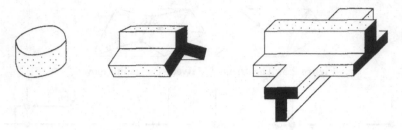

Figure 2.6: The three types of pieces by which manifolds are constructed

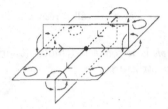

Figure 2.7: The orientation for the union of the 4 germs of disc (the horizontal disc in the figure) determines an embedding near the vertex

in four T's (two of them the right way up, and two upside-down). Every edge will prescribe how to (abstractly) glue together the T's corresponding to its ends, in a way which depends on its colour α. Namely, think of the edge as a segment drawn in front of you, and number the legs of the T's involved as shown in Figure 2.5 (right) depending on whether they are the right way up or upside-down. Then glue the T's matching the i-th leg on the right with the $(1,2,3)^\alpha(i)$-th one on the left. Having done all the gluings we have a neighbourhood of the singular set of a standard polyhedron. In other words, we only have to paste discs in the obvious way to get a standard polyhedron P. We define an orientation on P along the edges so that in the portions of Figure 2.5 we have the left-handed screw of $\mathbb{R}^3$. P determines a 3-manifold M just by "thickening" (M will be a regular neighbourhood of P in M itself), and M is oriented so that the screws of P are left-handed in M. One way to view the thickening process is as a puzzle game played with the three types of pieces shown in Figure 2.6: the matching instructions for the second and third types of pieces are carried by the o-graph, and then the pieces of the first type can be glued in without problems, because there is no Möbius strip on the boundary of the union of the pieces of the second and third type. This last fact is due to the orientation hypothesis and is the key step in the proof of Proposition 2.1.2.

FROM A 3-MANIFOLD TO SOME O-GRAPH. Given M let P be a standard spine of it, with at least two vertices. Orient P by the left-handed screw rule in M. Cut out an open disc inside each disc of P, to get a neighbourhood U of $S(P)$. Let v be a vertex of P. Choose at v a pair of opposite germs of disc, and remark that the union of the other four germs of disc can be regarded as a small disc D centred at v. Choose an orientation for D. Then there exists an essentially unique embedding in $\mathbb{R}^3$ of a neighbourhood of v in U such that: (1) D lies on the horizontal plane and its orientation matches the usual one of the horizontal plane; (2) the local orientation of P matches the orientation of $\mathbb{R}^3$ (i.e. the screws are left-handed: see Figure 2.7). Do the same at every vertex, to get embeddings in $\mathbb{R}^3$ of portions of U as described in

Figure 2.7, and assume that these embeddings are all at the same height and far from each other. Extend this partial embedding to the whole of U, in such a way that the projection of the singular set on the horizontal plane is generic. This same projection will be the support of resulting o-graph. Marked vertices will be the crossings which come from singularities of P (not the double points of the projection). The choice of under-over at marked vertices and of the colour to put on edges is easily obtained by reading backwards the procedure previously described (Figure 2.5).

Given these constructions the proof of Theorem 2.1.5 is obtained by taking into account the arbitrariness of the choices made to translate a spine into an o-graph, and Proposition 2.1.3. Another way to state Theorem 2.1.5, which we will use below, is the following:

Proposition 2.1.6. *Oriented standard spines correspond bijectively to o-graphs up to C, and MP translates the oriented Matveev-Piergallini move in the setting of o-graphs.*

Remark 2.1.7. The restriction in Theorem 2.1.5 that o-graphs should have at least two vertices allows to use, together with the C-move which is inevitably carried by the embedding in the plane, the MP-move only. On the other hand one could add another move which applies to o-graphs with one vertex and drop the restriction. One such a move will be shown much later in Figure 1.4. Note that the constructions outlined above are perfectly valid also when there is one vertex only.

2.2 Reconstruction of the boundary

In this section we recall from [43] (see also [44]) how to compute the boundary of the manifold associated to an o-graph. This will be needed in Chapter 3 for a combinatorial description of oriented branchings, and in Chapter 6 for the specialization to closed manifolds.

We first quickly introduce a representation for (oriented) surfaces by planar graphs. Let Θ be a trivalent graph in which each vertex has an assigned embedding in the plane, and each edge has a colour in $\mathbb{Z}_2$. We obtain a surface from such a Θ by first constructing a trivalent ribbon as shown in Figure 2.8 and then attaching a disc to every boundary component of the ribbon. If all the colours are 0 (and hence omitted) the resulting surface is oriented by the requirement that the portions shown in Figure 2.8 inherit the orientation of the plane.

Figure 2.8: How to construct the trivalent ribbon

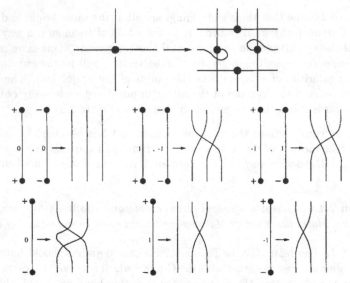

Figure 2.9: Boundary of an o-graph

Proposition 2.2.1. *Let Γ be an o-graph representing a 3-manifold M and let Θ be the graph associated to Γ as in Figure 2.9. Then the oriented surface associated to Θ as described above is the boundary of M.*

Proof of 2.2.1. The standard spine defined by Γ and the way M is reconstructed from it (the latter of which we are assuming to be known to the reader) naturally induce a decomposition of ∂M into pieces of 3 types: T-shaped strips, strips, discs; the first two types of pieces are shown in Figures 2.10 and 2.11, while discs appear as upper and lower caps of the thick discs $D^2 \times [-1, 1]$ glued in to obtain M. The same Figures 2.10 and 2.11 (together with the discussion of the remaining cases, which we leave to the reader) prove that ∂M is represented by the graph Θ_0 which is obtained from Γ via the rules of Figure 2.12. Now we must take into account the fact that in Figure 2.10 (and its translation in Figure 2.9) part of the boundary has been embedded with the wrong orientation. We can correct the orientation using the move of Figure 2.13, as shown on the right-hand side of the same figure.

It is now easy to check that the result is the graph Θ with all colours 0 (no colours) described in the statement. $\boxed{2.2.1}$

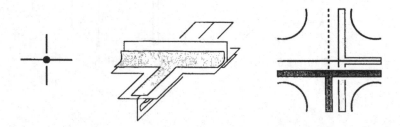

Figure 2.10: T-shaped strips on the boundary

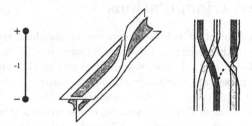

Figure 2.11: Strips on the boundary

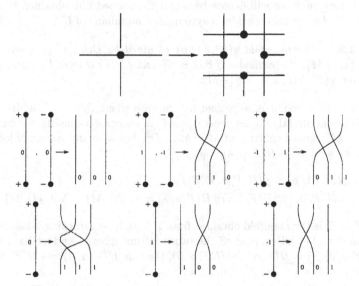

Figure 2.12: A non-oriented representation of the boundary

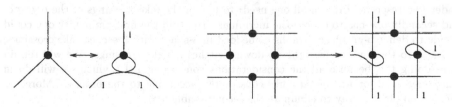

Figure 2.13: Correcting orientation

2.3   Surgery presentation of a mirrored manifold and ideal triangulations

In this section we will mention two facts concerning standard spines and o-graphs which are interesting in themselves and will also be used in Chapter 8. We first recall an explicit construction due to J. Roberts, to be found in [7], which takes from an o-graph to a surgery presentation of the manifold associated to the o-graph, glued to minus itself along the identity map on the boundary.

Let us consider an o-graph Γ with n vertices, representing a standard spine P of a 3-manifold M. We loosen the fake vertices of Γ in $\mathbb{R}^3$, so that a regular neighbourhood U of Γ in $\mathbb{R}^3$ is a handlebody of genus $g = n + 1$. One might also require that U is unknotted, but this is not essential under any respect. Now, the system of attaching circles for the discs of P can be viewed as a set δ of curves on $F = \partial U$. In the sequel all curves on F will be endowed with the framing induced by F; moreover if α and β are systems of curves on F we will denote by $\alpha \sqcup \beta$ the framed link obtained by pushing β slightly off U. Let us also consider a system μ of meridians of U.

Theorem 2.3.1. *The manifold $\chi(S^3, \delta \sqcup \mu)$ obtained by Dehn surgery on S^3 along $\delta \sqcup \mu$ is $M \sqcup_\partial (-M)$. In particular if $\partial M = S^2$ and $\widehat{M}$ is obtained by capping off the boundary then $\chi(S^3, \delta \sqcup \mu)$ is $\widehat{M} \# (-\widehat{M})$.*

Proof of 2.3.1. We provide an argument for the case where $\partial M = S^2$, addressing the reader to [7] for the slightly harder general proof. In this case δ consists of g curves and (U, δ) defines a Heegaard splitting of $\widehat{M} = M \sqcup_\partial D^3$. Let λ be any system of longitudes of F. Then we have $S^3 = H(F, \mu, \lambda)$ and:

$$\chi(S^3, \delta \sqcup \mu) = \chi(H(F, \mu, \lambda), \delta \sqcup \mu)$$
$$= H(F, \mu, \delta) \# H(F, \delta, \mu) \# H(F, \mu, \lambda) = \widehat{M} \# (-\widehat{M}) = M \sqcup_\partial (-M)$$

where $H(F, \alpha, \beta)$ is the manifold obtained from $F \times [0, 1]$ by attaching 2-handles along $\alpha \times \{0\}$ and $\beta \times \{1\}$ and capping off the two resulting spheres, and we have used the rule $\chi(H(F, \alpha, \beta), \gamma) = H(F, \alpha, \gamma) \# H(F, \gamma, \beta)$, the rule $H(F, \alpha, \beta) = -H(F, \beta, \alpha)$ and the facts $H(F, \mu, \delta) = \widehat{M}$ and $H(F, \mu, \lambda) = S^3$. $\boxed{2.3.1}$

Definition 2.3.2. *We will denote by $L(\Gamma)$ and call a Roberts link corresponding to Γ any framed link of the form $\delta \sqcup \mu$ obtained according to the process described above.*

Note that the process of constructing a framed link $\delta \sqcup \mu$ as above is ambiguous under two respects. First of all one needs to loosen the fake crossings of the o-graph, and secondly one has to choose the meridians. To avoid the first ambiguity one could decide to define o-graphs as graphs embedded in 3-space with loosened fake crossings, and add to the list of moves the up-down switches at fake crossings. We will refrain from doing this because all the considerations concerning $L(\Gamma)$ which we will do in Chapter 8 hold for *any* of the links constructed according to the process. Moreover there is no obvious way to eliminate the second ambiguity.

Remark 2.3.3. In [7] we show how to explicitly obtain a planar diagram for $L(\Gamma)$ starting from Γ, the most delicate point being the determination the framing. We will see later (see Chapter 8) that the theory of branched standard spines allows to take into account only a restricted class of o-graphs for which the framing can be determined in a much easier way. Therefore we do not recall here the general construction starting from an arbitrary o-graph.

Remark 2.3.4. Theorem 2.3.1 establishes a simple relation between the presentation of manifolds by means of standard spines (or more in general Heegaard splittings) and the surgery presentation. Note however that starting from a spine of M we obtain a surgery presentation of $M \sqcup_\partial (-M)$, and of course we lose some information in this passage (for instance $M \sqcup_\partial (-M)$ is always isomorphic to its mirror image, even if M is not). In [43] an algorithm has been exhibited which allows to obtain an o-graph presentation starting from any surgery presentation; this algorithm is quite complicated but it is perfectly practicable, in particular because it is linear with respect to some very natural complexity measures. On the other hand, to our knowledge, no practicable algorithm is known to translate a standard spine (or a triangulation) into a surgery presentation (Rourke's proof [50] that $\Omega_3 = 0$ is in principle constructive, but it is not clear to us how to translate it into an actual algorithm, and moreover there is very little obvious control on complexity). This problem will be referred to also in Chapter 9 (Questions 9.2.1 to 9.2.3).

Now we will recall the duality (see e.g. [38], [43]) between standard spines (and hence o-graphs) and ideal triangulations. This correspondence shows the relation with another fundamental area of research in 3-dimensional topology, namely hyperbolic geometry, concerning which we address the reader to [8], [38],[44], [58].

Let us first recall that, roughly speaking, an ideal triangulation of an oriented compact 3-manifold with non-empty boundary is a realization of the interior of the manifold as a finite union of tetrahedra with vertices removed and faces glued in pairs via orientation-reversing simplicial maps. One can show that there is a natural bijection between ideal triangulations up to combinatorial equivalence and standard spines up to homeomorphism. This bijection is induced by the duality suggested by Figure 2.14 (see [43] for details).

Rather than formalizing this process we draw a consequence which will be mentioned in Chapter 8 in connection with the Turaev-Viro invariants.

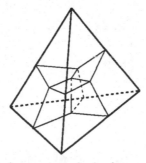

Figure 2.14: Duality between standard spines and ideal triangulations

Proposition 2.3.5. *A standard spine of a once-punctured closed 3-manifold M defines a cellularization of M with only one 0-cell, where each cell is a simplex of the appropriate dimension and its faces are glued to lower-dimensional cells via simplicial maps.*

Note that such a cellularization is very close to being a triangulation: the only difference is that there could be self-adjacencies of simplices and multiple adjacencies between different simplices (actually, certainly there are, because there is only one vertex).

Chapter 3

Branched standard spines

In this chapter we define the enhancement of the notion of standard spine by means of the concept of (oriented) branching. The description in this chapter is a purely combinatorial one, whereas starting from the next chapter we will analyze the topological counterpart on manifolds of the structures introduced here. After proving the equivalence of various different definitions of branched spine and introducing a very simple graphic encoding (a refinement of the notion of o-graph), we establish the main result of the chapter, according to which one can algorithmically find a branched spine of any manifold starting from an arbitrary standard spine of that manifold.

3.1 Branchings on standard spines

For the sake of brevity, from now on we will use the term standard spine to refer to an *oriented* one. The further structure we want to introduce on a standard spine is a coherent definition of a tangent plane for every point. In other words we would like to smoothen the corners of the spine, as intuitively shown in Figure 3.1. Since the idea is to have an analogue of a C^1-structure without boundary, it is quite natural to require that every C^1-function defined on $[0, 1]$ with values in the spine should have a C^1-extension on $\mathbb{R}$: this amounts to ruling out the situation shown in the right part of Figure 3.1. To conform our terminology with the literature, we will henceforth use the word *branching* to refer to such a C^1-structure.

We will now formalize our intuition in a combinatorial definition of a branching. The special case of an oriented branching will actually be much simpler, and the reader might also take Corollary 3.1.7 as a definition and proceed from there.

Definition 3.1.1. Let P be a standard spine. A *branching* on $P \setminus V(P)$ is a map which assigns to every edge e of P one of the three germs of disc which are incident to

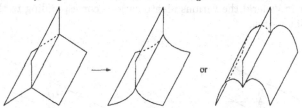

Figure 3.1: How to smoothen a spine (the second way is actually not acceptable for us)

e. The structure is said to *extend to* $v \in V(P)$ if the following happens. Let e_1, e_2, e_3, e_4 be the germs of edges incident to v, denote by $D_{\{i,j\}}$ the germ of disc at v incident to e_i and e_j, and let $D_{\{i,\alpha(i)\}}$ be the germ of disc which the structure associates with e_i: then α should attain 2 values, each of them twice. The structure is said to *extend to* $V(P)$ if it extends to every $v \in V(P)$.

We will now explain why this definition is natural. We first show how it allows to define an intrinsic tangent plane at every point of $P \setminus V(P)$, and then that the notion of extension to $V(P)$ is the natural one. Let P have m smooth components, denote by D the abstract disjoint union of m closed 2-discs, and let $\rho : D \to P$ describe how the discs are glued together to get the spine. Also, denote by ∂D the disjoint union of the boundary circles; D will be seen as a Riemannian 2-dimensional manifold with boundary, so every point x has attached a tangent plane $T_x D$ with scalar product, and for $x \in \partial D$ there is a line $T_x \partial D$ in this plane.

Fix an orientation for every component of $S(P) \setminus V(P)$ and orient $\partial D \setminus \rho^{-1}(V(P)) = \rho^{-1}(S(P) \setminus V(P))$ so that ρ preserves orientation. For $x \in \partial D \setminus \rho^{-1}(V(P))$ choose a basis $X(x), Y(x)$ of $T_x D$ as follows: $X(x)$ is the unit vector tangent to ∂D in the positive direction, and $Y(x)$ is the unit normal to $X(x)$ which points inside D. Let $\xi \in S(P) \setminus V(P)$ and remark that choosing a germ of disc incident to ξ corresponds to choosing one point in $\rho^{-1}(\xi) = \{x_1, x_2, x_3\}$.

Definition 3.1.2. Assume a branching is given on P, and assume that at ξ it corresponds to the choice of x_1. We define $T_\xi(P)$ as the disjoint union of the $T_{x_i} D$'s modulo the identification between them given as follows:

$$X(x_1) = X(x_2) = X(x_3), \qquad -Y(x_1) = Y(x_2) = Y(x_3).$$

Remark 3.1.3. This definition of $T_\xi P$ is actually independent of the orientation initially chosen for $S(P) \setminus V(P)$. In fact, changing the direction of the edge which contains ξ would result in simultaneously changing each $X(x_i)$ to $-X(x_i)$.

The geometric idea which underlies Definition 3.1.1 and the construction of tangent planes just given is explained in Figure 3.2. Now remark that if $v \in V(P)$ and e is a germ of edge at v, then e allows to single out 3 of the 6 points of which $\rho^{-1}(v)$ consists, call them x_1, x_2, x_3. Then we can define $X(x_i)$ and $Y(x_i)$ by taking the limit along $\rho^{-1}(e)$, and hence with the same rules as above we will have identifications between the $T_{x_i} D$'s. However in general the various identifications corresponding to the 4 germs of edges at v will not match.

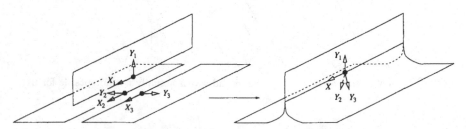

Figure 3.2: Construction of tangent planes

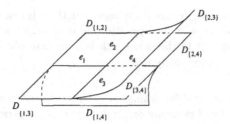

Figure 3.3: Condition of extension to a vertex

Proposition 3.1.4. *The branching on $S(P) \setminus V(P)$ extends to v according to Defini-tion 3.1.1 if and only if the identifications in $T_{\rho^{-1}(v)}D$ corresponding to the 4 germs of edge at v match with each other.*

Proof of 3.1.4. Fix the notation of Definition 3.1.1. Since $\alpha(i) \neq i$ the extension condition is that, up to renumbering the edges, α has the form:

$$\alpha(1) = 2, \quad \alpha(2) = 1, \quad \alpha(3) = 1, \quad \alpha(4) = 2.$$

Now, the fact that the identifications match means that a tangent plane is well-defined in a neighbourhood of v. This is equivalent to the fact that the neighbourhood can be C^1-identified (in an obvious sense) with the object shown in Figure 3.3. From the same figure one sees that the condition on α exactly translates the geometric compatibility condition.   3.1.4

As a consequence of this result the vertex-extension condition in Definition 3.1.1 can be expressed also in the following way: if we consider the chosen germ of disc as being the *preferred* one along the edge, then at each vertex one of the germs of disc should be preferred for both its edges, and the opposite germ of disc should be non-preferred for both its edges.

Definition 3.1.5. Fix the notation of Definition 3.1.1. An *orientation* for a branching on $P \setminus V(P)$ is an orientation for D (i.e. an orientation for each of the discs of P) such that all the identifications between the tangent spaces induced by the branching are orientation-preserving.

The following fact, though elementary, is quite surprising and pleasant. It is a manifestation of what Gillman and Rolfsen [20] call the circulation lemma.

Proposition 3.1.6. *If a branching on $P \setminus V(P)$ is orientable then it extends to $V(P)$*

Proof of 3.1.6. Let us identify a neighbourhood of $v \in V(P)$ with the cone based on the 1-skeleton of a tetrahedron. The cone on one side of the tetrahedron defines a germ of disc at v, and this is oriented by assumption, hence we can orient the side so that it leaves v on the left. Along an edge of the spine the 3 tangent planes to the discs are identified in an oriented way, and so that they do not give rise to "corners" (see Figure 3.1). In terms of the tetrahedron this means that no vertex is the source (or the target) of all 3 sides incident to it.

Let us say that a side of the tetrahedron is *alone* at the vertex w if either it points towards w and the other two edges incident to w point away from w, or the other way round. Under our condition one easily sees that there is exactly one side which is alone at both its endpoints, and the opposite side is not-alone at both its endpoints. Then the disc corresponding to the side which is alone will play the role of $D_{\{1,2\}}$ in Figure 3.3, and the conclusion easily follows. $\boxed{3.1.6}$

Corollary 3.1.7. *An oriented branching on P is an orientation for the discs of P such that no edge is induced the same orientation from all 3 germs of disc incident to it.*

From now on the notion of oriented branching which we will use in practice is that given by this corollary (but occasionally we will refer again to the underlying geometric idea). If P is a standard spine we denote by $C_i(P)$ the set of i-cells in P, each endowed with a fixed orientation. Having in mind the usual terminology in cellular $\mathbb{Z}$-homology, Corollary 3.1.7 leads us to consider for $i = 1, 2$ the set $\mathcal{F}_i(P)$ of i-chains in P which are *fundamental*, i.e. have all the coefficients equal to ± 1. Then the set of oriented branchings on P is given by:

$$\mathcal{B}(P) = \{\gamma \in \mathcal{F}_2(P) : \ \partial\gamma \in \mathcal{F}_1(P)\}.$$

This is exactly the object already considered in [20] and [21].

For the sake of brevity from now on we will write 'given $\beta \in \mathcal{B}(P)$' as a shorthand for 'given an oriented standard spine P with a certain oriented branching $\beta \in \mathcal{B}(P)$'. We will also call (P, β) (or P itself, if β is clear from the context) an oriented-branched standard spine, occasionally dropping the adjective standard.

3.2   Normal o-graphs

We prove in this section that $\beta \in \mathcal{B}(P)$ defines a canonical o-graph representation of P. Since $\partial\beta \in \mathcal{F}_1(P)$ the edges of P are naturally oriented, therefore so are the edges of an o-graph of P. Now let $v \in V(P)$. As shown in Figure 3.3 the branching near v allows to select 4 germs of disc at v. The union of these germs is a disc centred at v and the 4 orientations match. Therefore, by the very construction described in Section 2.1, we see that a definite o-graph corresponds to the oriented branching.

Proposition 3.2.1. *Let $\beta \in \mathcal{B}(P)$, and let Γ be the o-graph with oriented edges as just described. Then:*

1. *The orientations of the edges match across marked vertices.*

2. *Define the first and second end of an edge using the orientation. Then:*

    (a) If both ends are above or both are below, the colour of the edge is 0;

    (b) If the first end is below and the second end is above, the colour is 1;

    (c) If the first end is above and the second end is below, the colour is -1.

Conversely, if an o-graph Γ with these properties is given then on the standard spine defined by Γ there exists a unique oriented branching with which Γ is compatible.

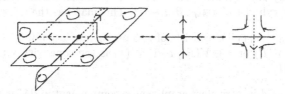

Figure 3.4: First possible local picture at a vertex

Figure 3.5: Second possible local picture at a vertex

Proof of 3.2.1. It is not difficult to show that essentially two local oriented branchings are supported by the portion of P which corresponds to a vertex of Γ (see Figures 2.5 and 2.7). These are shown in Figures 3.4 and 3.5, together with the corresponding local picture of Γ. The first assertion of the proposition follows at once from these pictures. For later purpose, on the right in these figures we are also using symbols introduced in [6], [43], which were not explicitly recalled above because their meaning should be clear anyway. Namely, the curves drawn represent, seen from above, the boundary of the portion of neighbourhood of the singular set which is shown on the left. In these particular figures the curves have also been oriented accordingly.

Now, for the second assertion, note that along every edge the colour prescribes a certain matching of the germs of disc, and this must respect the orientation. The only possible cases are those described in Figure 3.6, coherently with what was stated.

The inverse is proved just in the same way. Figures 3.4 to 3.6 allow to consistently define an orientation for the discs of the spine, and by construction on no edge the same orientation is induced 3 times. Uniqueness is obvious. $\boxed{3.2.1}$

According to this result, if we agree to represent every $\beta \in \mathcal{B}(P)$ by the corresponding o-graph Γ, we can also drop from Γ the colours of the edges, without losing any information. The intrinsic definition of normal o-graph was given in Chapter 1. A vertex of a normal o-graph (or the corresponding vertex of the associated branched spine) is said to have *index* +1 if it appears as in Figure 3.4, index −1 if as in Figure 3.5.

Remark 3.2.2. Since for normal o-graphs there is no more a C-move to take into account, we see that for them we have well-defined circuits, and moreover these circuits

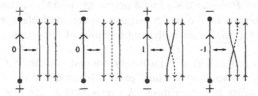

Figure 3.6: Admissible colours along edges

come with an orientation. We will see in the next section an intrinsic interpretation for these circuits.

3.3   Bicoloration of the boundary

Let P be a standard spine of an oriented 3-manifold with boundary M. Then there exists an essentially unique retraction $\pi : M \to P$ (induced by the collapse) such that M is the mapping cylinder of $\pi|_{\partial M} : \partial M \to P$. This map has the following properties:

1. For $x \in P$, $\pi^{-1}(x) \cap \partial M$ consists of 2, 3 or 4 points according as $x \in P \setminus S(P)$, $x \in S(P) \setminus V(P)$ or $x \in V(P)$, and $\pi^{-1}(x)$ is the cone with vertex x based on $\pi^{-1}(x) \cap \partial M$.

2. $\pi^{-1}(S(P)) \cap \partial M$ is a trivalent graph whose complement in ∂M is a disjoint union of discs each of which is homeomorphically mapped by π to some disc of P.

Let us fix one such $\pi : M \to P$. The following result provides a combinatorial description of oriented branchings on P by means of π. At the end of this section we will also consider the case of non-oriented branchings.

Proposition 3.3.1. *There exists a bijection between $\mathcal{B}(P)$ and the set of triples of the form (B, W, γ) such that:*

1. *∂M is the disjoint union of B, W and γ.*

2. *$\gamma \subset \pi^{-1}(S(P))$ is a disjoint union of circles.*

3. *Every component of γ has B on one side and W on the other side.*

4. *π maps $\gamma \setminus \pi^{-1}(V(P))$ bijectively onto $S(P) \setminus V(P)$.*

5. *For $v \in V(P)$, $\pi^{-1}(v) \cap \gamma$ consists of 2 points.*

6. *π maps B bijectively onto P, and the same for W.*

Remark 3.3.2.   1. The stated properties of the splitting $\partial M = B \cup W \cup \gamma$ are redundant, we have listed all of them because they give a precise description of the situation. We leave to the reader to find an equivalent smaller set of conditions.

2. As for property 3, recall that M is oriented so that ∂M is also oriented.

3. According to the proposition, the choice of $\beta \in \mathcal{B}(P)$ can be viewed as follows: for every edge of P choose one of its 3 preimages in ∂M, so that in ∂M the result is a union of loops whose complement can be coloured black and white, with no two regions of the same colour sharing portions of boundary.

Proof of 3.3.1. The proof is actually quite easy, we just outline it. Given $\beta \in \mathcal{B}(P)$ we paint black the (open) discs in ∂M which project in an orientation-preserving way to the discs of P, and white the other discs. A component of the preimage in ∂M of an (open) edge of P can separate 2 black discs, in which case we paint it black, or 2 white discs, in which case we paint it white, or a black and a white disc, in which case we leave it as it is. Similarly, if everything is black (white) near a point which projects to a vertex, we paint it black (white), otherwise we leave it as it is. The black and white regions B and W thus defined, together with the complement γ of their union, satisfy all the stated conditions.

Conversely, given a bicoloration of ∂M, we define an orientation on the discs of P by the requirement that π is orientation-preserving on the black discs of ∂M. $\boxed{3.3.1}$

The reader should keep in mind that a bicoloration is an *ordered* triple, namely that black and white regions are not interchangeable: the black region will always be the one which projects orientedly to the spine. Interchanging black and white regions would lead to the opposite oriented branching. In Section 4.1 we will see that the bicoloration associated to a branched standard spine may also be described in terms of a more elaborate structure.

Remark 3.3.3. We can now give an interpretation to the oriented circuits of an o-graph. In fact it is not difficult to see that they naturally correspond to the components of γ, oriented as boundary of the black region.

The black and white regions of a bicoloration of ∂M associated to an oriented branching of P are non-compact surfaces, which can be canonically compactified to surfaces with boundary by taking their closure in ∂M (in other words, adding γ). We do not introduce another term to name this compactification, it will be clear from the context if we refer to the compact or non-compact region.

Proposition 3.3.4. *The black region and the white one have the same Euler characteristic and the same number of boundary components.*

Proof of 3.3.4. We can compute χ using the decompositions into open cells of the non-compact regions. Since π is bijective onto P when restricted to either region, we see that χ is just $\chi(P) = \chi(M)$. The boundary components are exactly the components of γ, for both the regions. $\boxed{3.3.4}$

Let us describe explicitly how to obtain the bicoloration of the boundary from a normal o-graph. This will require the procedure for reconstructing the boundary given in Section 2.2.

Proposition 3.3.5. *Let Γ be a normal o-graph which defines $\beta \in \mathcal{B}(P)$ as in Proposition 3.2.1, and let P define a 3-manifold M. The bicoloration of ∂M associated to $\beta \in \mathcal{B}(P)$ is obtained as follows:*

1. *Produce an oriented bicolorated surface with boundary according to the rules of Figures 3.7 and 3.8. Remark that each boundary circle has a definite colour.*

2. *Glue discs to turn this surface into a closed oriented one, painting each disc the same colour as the circle it is attached to.*

Proof of 3.3.5. Looking at Figures 3.4 and 3.5, and thinking of a portion of the 3-manifold as a thickening of the portion of spine shown, one sees that most of the boundary is naturally smooth, but there are bending arcs carried by the branching, which are just the separating arcs between black and white region, as shown in the

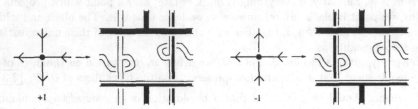

Figure 3.7: To every vertex of a normal o-graph, in a way which depends on the index, we associate a portion of bicolorated oriented surface with boundary (in the plane of this figure, all pieces of surface are drawn with their orientation)

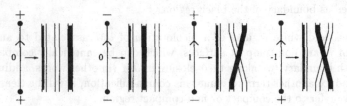

Figure 3.8: The edges of a normal o-graph prescribe how to connect the portions of surface produced above (this is done abstractly, so unmarked crossings do not affect the surface)

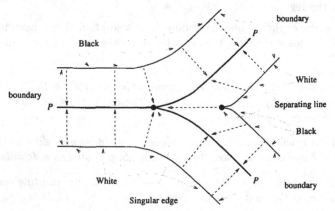

Figure 3.9: A local cross-section far from vertices. The dashed arrows represent the projection

cross-section of Figure 3.9. The trivalent ribbons which appear in Figures 3.7 and 3.8 represent portions of the boundary (but they are not drawn in the obvious way, because we want the pictured portions to have the right orientation: we address the reader to the proof of Proposition 2.2.1). And it is not difficult to see that the bicoloration (in particular, the position of the separating lines) is the correct translation of Figure 3.9. (The reader is invited to fill in the details we have skipped.) ⟦3.3.5⟧

We close this section with a brief digression concerning an alternative characterization of branchings, which is particularly suitable for the non-oriented case. The combinatorial definition 3.1.1 of non-oriented branchings is rather complicated, but the next result, due to Christy [12], implies that they can be encoded in a rather simple way on the boundary of the manifold. Let us fix the projection $\pi : M \to P$ of a manifold with boundary on a standard spine, as at the beginning of the section. Recall that $\pi^{-1}(S(P)) \cap \partial M$ is a trivalent graph in ∂M.

Proposition 3.3.6. *Branchings on P correspond in a natural and bijective way to subgraphs Δ of $\pi^{-1}(S(P)) \cap \partial M$ such that:*

 (*i*) *Δ consists of a disjoint union of circles;*

 (*ii*) *Δ contains exactly one of the three preimages of each open edge of $S(P)$.*

Proof of 3.3.6. We confine ourselves to a sketch, leaving the details to the reader. By definition a branching is the choice of a preferred germ of disc along each edge, with the following compatibility condition: at each vertex one of the germs of disc is preferred for both its edges, and the opposite germ of disc is non-preferred for both its edges. Now let Δ be a subgraph of $\pi^{-1}(S(P)) \cap \partial M$ as described in the statement. Let e be an open edge of $S(P)$ and let $\tilde{e}$ be the component of $\pi^{-1}(e)) \cap \partial M$ which lies in Δ. A neighbourhood of $\tilde{e}$ in ∂M will project onto two of the three germs of disc at e, and then we define the third one to be the preferred one. To check compatibility at vertices one uses the symmetries of a neighbourhood of the vertex to deduce that the combinatorial pattern is always as in Figure 3.3.

The process which associates a subgraph of $\pi^{-1}(S(P)) \cap \partial M$ to a branching is completely analogous. ⟦3.3.6⟧

To state an analogue of Proposition 3.3.6 for the case of oriented branchings let us first remark that any graph Δ with properties (*i*) and (*ii*) necessarily contains exactly two preimages of each vertex. If the edges of $\pi^{-1}(S(P)) \cap \partial M$ are oriented a vertex v of Δ is called incoming if the the germ of edge at v which does not belong to Δ points towards v. We omit the proof of the following result.

Proposition 3.3.7. *Oriented branchings on P correspond bijectively to subgraphs Δ of $\pi^{-1}(S(P)) \cap \partial M$ with properties (i), (ii) and:*

(*iii*) *the edges of $\pi^{-1}(S(P)) \cap \partial M$ can be oriented in such a way that:*

 (*a*) *the orientations are compatible under π;*

 (*b*) *Δ is naturally oriented;*

 (*c*) *exactly one of the preimages of each vertex of P is incoming for Δ.*

3.4　Examples and existence results

We give now some examples concerning the notions we have introduced and their independence, their relation with the Matveev-Piergallini move and with orientability of the spine. Later in this section we will prove general existence results.

Example 3.4.1. The spine shown on the left in Figure 3.10 does not admit any branching (not even non-orientable). The argument is a rather long case-by-case one: for all the $3^4 = 81$ possible choices along edges one checks that extension to vertices fails. Moreover we have shown in the same Figure 3.10 an MP-move, and subsequently C-moves which allow to obtain a normal o-graph. (Move C has been applied 3 times, 2 of which at the same vertex without showing the intermediate passage.) Then Proposition 3.2.1 implies that the new spine has an oriented branching. Therefore this is *a non-branchable spine which evolves via one move to an oriented-branched spine*. More will be said in the sequel on the behaviour of oriented branchings under MP.

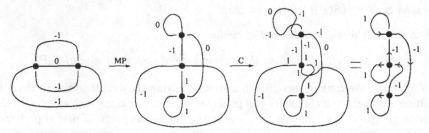

Figure 3.10: A non-branchable spine which evolves into an oriented-branched spine

Example 3.4.2. In Figure 3.11 we describe *a spine which admits branchings but not oriented branchings*. In the figure we have described a branching by placing small squares on the chosen germs of disc at the edges. It is an easy exercise to prove that this branching indeed extends to the vertex. Now the only o-graphs which are obtained by C-moves from the original one are those shown in the same picture. Since they cannot be oriented so to become normal o-graphs, it follows that the spine admits no oriented branching.

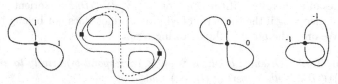

Figure 3.11: A branched spine which is not oriented-branchable

Example 3.4.3. In Figure 3.12 we show a normal o-graph and a branching on the spine represented by it, and leave to the reader to check that such a branching is not orientable. So this is *a spine which has both orientable and non-orientable branchings*.

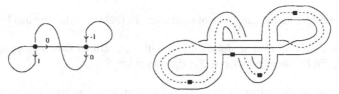

Figure 3.12: An oriented and a non-orientable branching on the same spine

Example 3.4.4. Let us forget for a moment our convention that standard spines are automatically oriented. Having in mind Corollary 3.1.7 one could define also in this case an oriented branching as an orientation for the discs which behaves well along the edges. Then one may wonder if the existence of an oriented branching forces a spine to be orientable. As already noted in [21] this is not the case. Figure 3.13 shows the neighbourhood of the singular set of a non-orientable standard polyhedron, which can be shown to be a spine of a punctured $\mathbb{P}^2 \times S^1$, and the figure also shows the oriented branching (since this example is not central to us we skip details on representation of non-orientable spines).

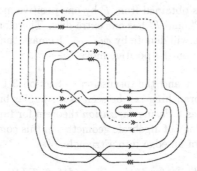

Figure 3.13: An oriented branching on a non-orientable spine: the two small circles correspond to regions where there is no actual intersection

Example 3.4.5. We provide here an example of bicoloration of the boundary where *white and black region are not homeomorphic* (recall that they always have the same Euler characteristic and the same number of boundary components). Consider the normal o-graph given in Figure 3.14. As a direct application of Proposition 3.3.5 one sees that the boundary of the associated manifold is a torus and that the black region on it is an annulus whose core is homotopically trivial. In particular the black region is connected while the white one has two components.

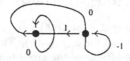

Figure 3.14: A normal o-graph

We start facing general questions of existence. In [20] the following has been shown:

Proposition 3.4.6. *If P is a standard spine of a punctured $\mathbb{Z}$-homology sphere then for every $\delta \in \mathcal{F}_1(P)$ there exists $\beta \in \mathcal{F}_2(P)$ with $\partial\beta = \delta$.*

Therefore we deduce from Corollary 3.1.7 that every spine of a punctured $\mathbb{Z}$-homology sphere admits oriented branchings. We also have:

Proposition 3.4.7. *Every oriented three-dimensional manifold with boundary admits a standard spine with an oriented branching.*

This result by itself is not very difficult and was already remarked in [21], even if not strictly in our context (Gillman and Rolfsen deal with closed manifolds only, and they accept simple spines). We will deduce Proposition 3.4.7 by considering the relation with the MP-move and proving a more specific result which has independent interest.

In the sequel, to emphasize that we are performing a move in the direction which increases the number of vertices, we will call it a *positive* MP-move.

Definition 3.4.8. If P' is obtained from P by one positive move, every disc in P can be identified to a disc in P', and P' has one extra disc. Therefore we have a 2-to-1 map $\mathcal{F}_2(P') \to \mathcal{F}_2(P)$ which we will denote by $\phi_{(P \to P')}$. Remark that this map depends not only on P and P' but also on the positive move $P \to P'$.

According to Figure 3.1 and the discussion of Section 3.1 a fundamental 2-chain which is not an oriented branching can be seen as a singular branching. Therefore the next result could be seen as a *desingularization* theorem for fundamental 2-chains. By analogy with the terminology of algebraic geometry, in this context one could view the MP-move as the *blow-up* of an edge into a triangle.

Theorem 3.4.9. *Let P be a standard spine and $f \in \mathcal{F}_2(P)$. Then there exist a sequence of positive MP-moves $P = P_0 \to P_1 \to \cdots \to P_m$ and a branching $c \in \mathcal{B}(P_m)$ such that $f = \left(\phi_{(P_0 \to P_1)} \circ \cdots \circ \phi_{(P_{m-1} \to P_m)}\right)(c)$.*

Since we know that the MP-move relates spines of the same manifold, Theorem 3.4.9 implies Proposition 3.4.7. In the proof we will also see that the sequence of moves in Theorem 3.4.9 is determined algorithmically by the initial data, and that one can always choose $m \leq 5n$, where n is the number of vertices of P.

Proof of 3.4.9. For $\gamma \in \mathcal{F}_2(Q)$, an edge $e \in C_1(Q)$ is called *bad* if it has coefficient ± 3 in $\partial\gamma$, *good* otherwise. Let $Q \to Q'$ be the positive move which blows up an edge e_0 and causes edges e_1, e_2, e_3 to appear. Then we have a natural bijection:

$$C_1(Q') \setminus \{e_1, e_2, e_3\} \to C_1(Q) \setminus \{e_0\}.$$

Moreover for $\gamma \in \mathcal{F}_2(Q')$ we have that $e \in C_1(Q') \setminus \{e_1, e_2, e_3\}$ is bad for γ if and only if the corresponding edge of Q is bad for $\phi_{(Q \to Q')}(\gamma)$. The idea of the proof is to blow up bad edges and choose suitable liftings of the given fundamental 2-chain so that the total number of bad edges decreases. This idea must however be refined in particular to deal with bad loops.

Figure 3.15: Proof of facts 1 and 2

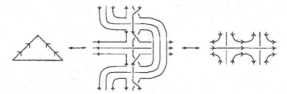

Figure 3.16: Proof of fact 2 continued

We prove various facts which will lead to the conclusion. In the figures we always freely use the correspondence between spines and o-graphs up to C-move.

FACT 1. *For $\gamma \in \mathcal{F}_2(Q)$ and $v \in V(Q)$, at most 2 germs of edge at v are bad.* If 2 edges are bad, without loss of generality the situation is as in Figure 3.15 (left), and the conclusion is obvious.

FACT 2. *Let $\gamma \in \mathcal{F}_2(Q)$ and $e_0 \in C_1(Q)$. Assume e_0 is bad, has distinct endpoints and not at both endpoints there is another bad edge. If $Q \to Q'$ is the blow-up of e_0 then there exists $\gamma' \in \phi_{(Q \to Q')}^{-1}(\gamma)$ such that (Q', γ') has one less bad edge than (Q, γ).* Without loss of generality the situation is as in Figure 3.15 (right). Consider the triangle which is originated by the move. If we lift γ we have 2 orientations on each edge of the triangle. Assume by contradiction that whatever orientation we put on the triangle, one edge will always be bad. Then there exists either one vertex of the triangle which is a "quadruple source" or one which is a "quadruple target". Up to reversing all orientations we can assume the latter case. Then, using the symmetry of the move, we can assume the situation is as in Figure 3.16, where we also show the corresponding situation before the move, which gives a contradiction because, whatever choice of orientation on the 2 unoriented discs, at both ends of the edge there are 2 bad germs.

FACT 3. *Let $\gamma \in \mathcal{F}_2(Q)$ and $e_0 \in C_1(Q)$. Assume e_0 is bad, has distinct endpoints and at both endpoints there is another bad edge. If $Q \to Q'$ is the blow-up of e_0 then either there exists $\gamma' \in \phi_{(Q \to Q')}^{-1}(\gamma)$ such that (Q', γ') has one less bad edge than (Q, γ), or there exists another move $Q' \to Q''$ and $\gamma'' \in (\phi_{(Q \to Q')} \circ \phi_{(Q' \to Q'')})^{-1}(\gamma)$ such that (Q'', γ'') has one less bad edge than (Q, γ).* We have to make do with the situation of Figure 3.16. There are 4 different possibilities to consider for the orientation, and for each of them we have shown in Figure 3.17 a choice of orientation for the triangle, and which edge to blow up next. In the second and fourth case Fact 2 directly implies that there is a lift with less bad edges. In the first and third case one must show that, even if there are indeed other bad edges at both ends, the combinatorial situation is not equivalent to that of Figure 3.16. This is not difficult and left to the reader.

FACT 4. *Let $\gamma \in \mathcal{F}_2(Q)$, let $e \in C_1(Q)$ be a bad loop at $v \in V(Q)$, let $e_0 \in C_1(Q)$ have endpoint v and let $Q \to Q'$ be the blow-up of e_0. Assume that there is no $\gamma' \in \phi_{(Q \to Q')}^{-1}(\gamma)$ such that (Q', γ') has as many bad edges as (Q, γ) and one less bad loop. Further, let*

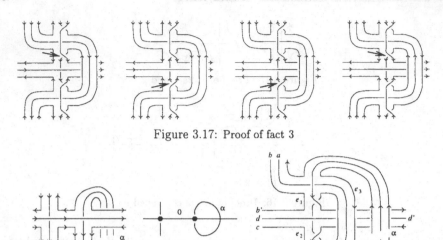

Figure 3.17: Proof of fact 3

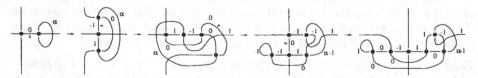

Figure 3.18: Statement and proof of fact 4

$e_1, e_2, e_3 \in C_1(Q')$ be the edges originated by $Q \to Q'$, and let $Q' \to_i Q_i$ be the blow-up of e_i. Assume that for $i = 1, 2, 3$ there is no $\gamma'' \in (\phi_{(Q \to Q')} \circ \phi_{(Q' \to_i Q_i)})^{-1}(\gamma)$ such that (Q_i, γ'') has as many bad edges as (Q, γ) and one less bad loop. Then without loss of generality the situation is as is Figure 3.18 (left). Of course we can use the o-graph shown in the middle of Figure 3.18, and assume the directions on the loop are as shown. After the blow-up we have the situation shown on the right in the same picture. The first condition implies that the directions must be from a to a', and either from b to b' or from c to c' (or both). Now, if for instance the direction is from c' to c then we can orient the new disc so that only e_1 is bad, and by Fact 2 we see that $Q' \to_1 Q_1$ contradicts the second assumption. Similarly we will have necessarily a direction from b to b'. Now it might well happen that the direction is not from d to d', but then we can simultaneously reverse all directions and take the mirror of the figure in a line of the plane. This will not affect our combinatorial considerations, and allows us to assume that the direction is from d to d'.

FACT 5. *In the situation of Figure 3.18 (left) there exist a sequence of blow-ups $Q \to P_1 \to P_2 \to P_3 \to P_4 = Q'$ and $\gamma' \in \mathcal{F}_2(Q')$ which projects onto γ and such that (Q', γ') has as many bad edges as (Q, γ) and one less bad loop.* In Figure 3.19 we show the sequence of moves (we do not make explicit all the C-moves and isotopy which are used, we only indicate by small dotted arrows the edges which are blown up). Figure 3.20 displays the fundamental 2-chain γ' which proves Fact 5, because in the picture there

Figure 3.19: Proof of fact 5

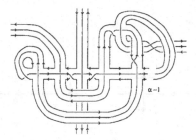

Figure 3.20: Proof of fact 5 continued

is only one bad edge, and it is not a loop.

CONCLUSION. By induction on the number of bad edges. If some bad edge is not a loop, we use facts 2 and 3, otherwise we use fact 5 to get a bad edge which is not a loop. (Remark that it can actually happen that removing a bad edge we introduce bad loops, so it is not possible in general to first remove the bad loops and then proceed removing bad edges). $\boxed{3.4.9}$

3.5 Matveev-Piergallini move on branched spines

Let us recall that on standard spines we have a move called MP which generates the calculus for 3-manifolds in presence of at least 2 vertices. Having introduced oriented branchings as a further structure on standard spines, in this section we investigate the natural question of how does this structure behave in connection with the MP-move. The arguments in this section only deal with this natural purely combinatorial problem, without 3-manifold interpretations; however we will see in Section 4.3 that the results do have a topological interpretation. From now on branchings will always be oriented.

We have already seen in Example 3.4.1 that there are negative MP-moves on branched spines whose outcome cannot be given a branching. The next result exactly describes the situations when this happens, and, as opposed to this, implies that the outcome of a positive MP-move on a branched spine is always branchable, but not uniquely in some cases. The result is expressed in terms of normal o-graphs, and could be translated in terms of spines via Proposition 3.2.1.

Proposition 3.5.1. *Consider a normal o-graph and apply to it a positive MP-move. Then the result can always be represented by a normal o-graph, and the complete table of possibilities is given by Figures 3.21 and 3.22, where in Figure 3.22 the unoriented circuits after the move can be oriented arbitrarily. Moreover the moves of Figure 3.21 do not alter the bicoloration of the boundary given by Proposition 3.3.1, while the moves Figure 3.22 change it as in Figure 3.23, or a similar figure with colours interchanged.*

Remark 3.5.2. The cases where the outcome of the move can be turned into an o-graph in two different ways are not only those of Figure 3.22. In fact, if an o-graph matches both a pattern of the first row of Figure 3.21 and the same pattern rotated in the plane, we can apply to it the move in two different ways (see also the first line of Figure 1.8, ignoring the $\mathbb{Z}_2$-coloring).

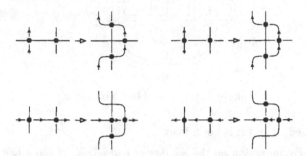

Figure 3.21: Moves on normal o-graphs (these are exactly the same as in Figure 1.5; they have been repeated to have the complete list at hand, and because here they only have a combinatorial motivation, whereas in Figure 1.5 they are part of a geometric argument

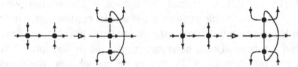

Figure 3.22: More moves on normal o-graphs

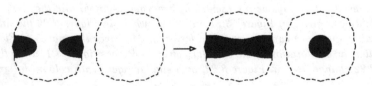

Figure 3.23: Change of bicoloration

Remark 3.5.3. In the language of fundamental classes (see the end of Section 3.1), we know that under a positive move $P \to P'$ an element of $\mathcal{F}_2(P)$ lifts to 2 elements of $\mathcal{F}_2(P')$. Then the above proposition implies that every element of $\mathcal{B}(P) \subset \mathcal{F}_2(P)$ lifts to either 1 or 2 elements of $\mathcal{B}_2(P') \subset \mathcal{F}_2(P')$, and both cases occur.

Remark 3.5.4. It can be checked by direct inspection that in each of the moves of Theorem 3.5.1 the sum of the indices of vertices changes by ± 1.

Proof of 3.5.1. As we said above, this is just a combinatorial argument without topological interpretation, so we just outline it without figures. One starts with any edge with different ends in a normal o-graph (there are 16 different possible patterns), and applies to it C-moves so to obtain an o-graph to which the MP-move applies (Figure 2.4). Then one only has to determine which combinations of C-moves on the outcome are compatible with a normal o-graph (with the previously existing orientations being preserved). Note that a priori 6 moves should be considered for each of the 3 vertices, in all 16 cases, but actually there are shortcuts and the whole argument can be rather quickly carried out by hand. It is also not difficult to summarize the result as represented in Figures 3.21 and 3.22.

The evolution of the bicoloration under the moves is also a not very difficult application of Proposition 3.3.5. Note also that Remark 3.3.3 itself is already sufficient to see that the circuits which give the bicoloration are unchanged under the moves of Figure 3.21, while under the moves of Figure 3.22 two adjacent circuits are connected and a new circuit is born somewhere else. $\boxed{3.5.1}$

Chapter 4

Manifolds with boundary

In this chapter we will describe additional structures on 3-manifold associated to oriented-branched standard spines. Namely we will establish various relations between oriented branchings and non-singular flows with special properties on the boundary (up to a suitable equivalence relation). This result will naturally lead us to to a slight refinement of the notion of standard spine (Definition 4.1.12): under this refinement a spine will determine a manifold in a more precise way than up to homeomorphism. Moreover we will introduce an equivalence relation on flows (a special type of homotopy) and determine the combinatorial moves on oriented-branched standard spines which translate this equivalence relation. In the last section we will consider vector fields up to homotopy and prove some related results.

4.1 Oriented branchings and flows

This section is devoted to establishing the topological counterpart on manifolds of the notion of branching on standard spines. Our construction is partially inspired by Ishii's ideas [24], [25]. We refer to Chapter 5 for a more detailed account of these works and the relations of our results with those of [26].

Since flows are well-defined only in a differentiable setting, we will need in this section some care about how to smoothen topological 3-manifolds. The main technical point is given by the next lemma and the subsequent remark. If M is a smooth 3-manifold with boundary, let us call *smooth generic flow* on M a smooth nowhere-vanishing vector field X on M such that X is tangent to ∂M exactly at the points of a compact 1-dimensional submanifold $\gamma \subset \partial M$, and X is never tangent to γ.

Lemma 4.1.1. *Let M, X, γ be as above. Then every point p of γ has a neighbourhood U in M such that $(U, X|_U, p, \gamma \cap U)$ is equivalent (via a diffeomorphism) to*

$$\big(\{(x, y, z) \in \mathbb{R}^3 : -\varepsilon < x, y < \varepsilon, f(x) \le z < \delta\}, \partial/\partial x, 0, \{0\} \times (-\varepsilon, \varepsilon) \times \{0\}\big),$$

where $\varepsilon, \delta > 0$, $f(0) = f'(0) = 0$ and f is either convex or concave in $(-\varepsilon, \varepsilon)$. (See Figure 4.1 for a description.) We will then say that p is respectively a convex or concave point of the boundary with respect to X. No point is both convex and concave, and all the points of a component of γ are of the same type.

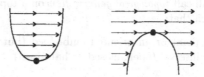

Figure 4.1: Convex and concave points on the boundary

Proof of 4.1.1. Locally let $M = \{(x, y, z) \in \mathbb{R}^3 : -\varepsilon < x, y < \varepsilon, h(x, y) \le z < \delta\}$, with $h(0) = (\partial h/\partial x)(0) = (\partial h/\partial y)(0) = 0$, $p = 0$ and $X(0) = \partial/\partial x$. Using the change of coordinates $(x, y, z) \mapsto \psi_{(y,z)}(x)$, where $\psi_{(y,z)}$ is the solution of:

$$\begin{cases} \psi_{(y,z)}(0) = (0, y, z) \\ (d\psi_{(y,z)}/dx)(x) = X(\psi_{(y,z)}(x)), \end{cases}$$

we see that we can assume $X \equiv \partial/\partial x$. So locally:

$$\gamma = \{(x, y, h(x, y)) : (\partial h/\partial x)(x, y) = 0\}.$$

Since we know that $\partial/\partial x$ is not tangent to γ in 0 we deduce that $(\partial^2 h/\partial x^2)(0) \ne 0$. For small y the function $x \mapsto h(x, y)$ will attain its only local minimum or maximum (according to the sign of $(\partial^2 h/\partial x^2)(0)$) at a point $x_*(y)$, and of course x_* is a smooth function. The change of coordinates

$$(x, y, z) \mapsto (x - x_*(y), y, z - h(x_*(y), y))$$

of course allows to meet the prescriptions of the statement. The final assertions are obvious. $\boxed{4.1.1}$

Remark 4.1.2. In the previous setting one may wonder if a special form for f, e.g. $f(x) = kx^2$, can be obtained. If for instance $f''(0) > 0$, we can define for $y > 0$

$$M(y) = \max\{x : f(x) = y\}, \quad m(y) = \min\{x : f(x) = y\}, \quad L(y) = M(x) - m(x)$$

and set

$$Y(y) = \begin{cases} y & \text{if } y \le 0 \\ f''(0) \cdot L(y)^2/8 & \text{if } y > 0 \end{cases}$$

$$X(x, y) = \begin{cases} x & \text{if } y \le 0 \\ x - M(y) + \sqrt{2Y(y)/f''(0)} & \text{if } y > 0 \end{cases}.$$

If we change coordinates using $(x, y) \mapsto (X, Y)$, we have indeed that the new form of f is $X \mapsto kX^2$; however the map $(x, y) \mapsto (X, Y)$ is not C^∞ in general (but for instance one sees that it is C^1 in y). A similar remark holds for the case $f''(0) < 0$.

Remark 4.1.3. Knowing a priori that (M, X) is a generic smooth flow one can a tell convex point on the boundary from a concave one just by looking at the integral curves of X, even without actually knowing X. This is because a convex point is itself a (degenerate) orbit, while a concave point is part of an orbit which near the point (except the point itself) lies in the interior.

Definition 4.1.4. We will call *topological generic flow* on a topological 3-manifold M a partition $\{A_i\}$ of M such that:

1. Each A_i is either a point or an oriented embedded (but possibly not properly embedded) circle or interval (open, closed or half-open);

2. There exist a smooth 3-manifold M_s, a smooth generic flow X on M_s and a homeomorphism f of M onto M_s such that $\{f(A_i)\}$ is the set of orbits of X, with correct orientation.

The subsets A_i will be referred to as *orbits* also in this topological setting. If a manifold is determined only up to (oriented) homeomorphism, two topological generic flows on one of its representatives are regarded as equivalent if they are mapped to each other by an oriented homeomorphism, so that the notion is well-defined also in this case.

Remark 4.1.5. In a smooth flow an internal point is always internal to an orbit, so the same holds in the topological case. Therefore, also for a topological flow, an endpoint of an orbit necessarily belongs to the boundary of the manifold, and using the orientation we can define if the flow points outside or inside the manifold at the point.

Remark 4.1.6. By Remark 4.1.3 if a homeomorphism maps the orbits of a smooth generic flow to those of another smooth generic flow, then it maps convex and concave points on the boundary to points of the same sort. Therefore convex and concave points are well-defined also for generic topological flows. Moreover, according to Lemma 4.1.1, concave points are exactly the points of the boundary which belong to the interior of some orbit, and convex points are exactly the points which are themselves orbits.

Using Remarks 4.1.5 and 4.1.6 we see that the boundary of a manifold with a generic topological flow splits into a *black* and a *white* region (respectively, the set of points where the flow points outside and inside) and a finite number of separating loops which are made of convex or concave points. From now on we will call just flow a topological flow, and occasionally refer to smooth versions of it. We need a few more definitions before stating the main result of this section.

Definition 4.1.7. Consider a generic smooth flow on a manifold M generated by a vector field X, and let γ be the separating curve on ∂M. Let two concave points $p_0, p_1 \in \gamma$ be joined by a segment of orbit. We will say that γ is *transversal* to itself across the segment if the following holds: consider the isomorphism of tangent spaces $\phi : T_{p_0} M \to T_{p_1} M$ obtained by integrating X; then $\phi(T_{p_0}\gamma)$, $T_{p_1}\gamma$ and $X(p_1)$ should span $T_{p_1} M$. (See below Figure 4.2-right for a geometric interpretation of this definition.)

Definition 4.1.8. A generic topological flow on a 3-manifold with boundary is called a *traversing flow* if all the orbits are homeomorphic to closed intervals or points, and *concave* if there are no orbits homeomorphic to points (in this case of course there are no convex points).

Theorem 4.1.9. *Let P be an oriented standard spine of a manifold M (up to homeomorphism). The rules of Figure 4.2 allow to associate to every oriented branching of P a concave traversing flow on M, with the further properties that:*

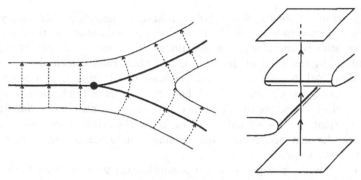

Figure 4.2: Flow associated to a branching. In the cross-section on the left the orientation of the discs is such that the flow is positively transversal to them. The figure on the right suggests the effect of a vertex

1. *Either an orbit is properly embedded or its interior intersects the boundary in exactly one or two (necessarily concave) points, and the orbits with two intersections are finitely many;*

2. *There exists a smoothing of the flow in which the set γ of concave points is transversal to itself across any orbit segment with both ends on γ.*

3. *Consider the trivalent graph T on ∂M which consists of γ and all the endpoints of the orbits which intersect γ; then the black (or, equivalently, white) components of $\partial M \setminus T$ should be discs.*

Moreover to every concave traversing flow on an oriented manifold with boundary satisfying these properties there corresponds a unique oriented-branched standard spine of the manifold which induces the given flow.

Proof of 4.1.9. Let us first note that Figure 4.2 gives a qualitative description of how the orbits should be, *only once the projection of M onto P has been fixed.* On the other hand we know that different projections are related by a self-homeomorphism of M, so recalling Definition 4.1.4 we see that we can stick to a given realization of P as spine of M. Then, we must show that:

A. There exists a smoothing of M and a smooth generic flow on it whose orbits agree with the qualitative description of Figure 4.2, and such that property 2 is satisfied;

B. Given two such smoothings, there exists a self-homeomorphism of M which maps the orbits of one to those of the other.

(This implies that the flow is well-defined, and then the other properties are obvious: only the last one requires a little reasoning.) To prove point A, we cut an oriented-branched standard spine into pieces much as is done in [6] (see also Figure 2.6), so that the spine can be seen as a union of finitely many building blocks of prescribed shape, and correspondingly the manifold is a union of blocks which thicken those of the spine. On the abstract models of the blocks we fix smooth flows which agree with the description of Figure 4.2. For well-chosen abstract models one can always realize the gluings of the pieces in a way which respects the smoothing and the flow.

To prove point B, consider the bicolorations of the boundary induced by two flows as in point A (in the fixed model of M), and their enhancement by the trivalent graphs as described in point 3. Since the combinatorial situation can be completely described using Figure 4.2, there exists a homeomorphism of the boundary isotopic to the identity which maps one enhanced bicoloration to the other one. Now to extend the homeomorphism inside we only need to follow the orbits, which we can do in a canonical way for instance fixing a Riemannian structure and making sure that we proceed at constant speed.

Given a flow with these properties, it is not difficult to reconstruct the spine: one just takes the closure of the black region and identifies points which belong to the same orbit. The smooth components of the spine are naturally identified with the black components of $\partial M \setminus T$, and they are given the corresponding orientation. $\boxed{4.1.9}$

Definition 4.1.10. Given a concave traversing flow, the curves along which the flow is tangent to the boundary can be seen as curves of apparent contour of the boundary with respect to the flow, and will be referred to as *contour curves*.

Remark 4.1.11. By comparing Figures 3.9 and 4.2 one easily sees that the bicoloration associated to the flow carried by a branching is the same as that introduced in Section 3.3.

As announced, we discuss now a slight refinement of the notion of standard spine which is naturally suggested by our construction of a flow associated to the spine. The point we want to make is the following: as noted along the proof of Theorem 4.1.9, the flow itself is very tightly determined by the spine once the position of the spine is fixed in a certain manifold, but without knowing this position we are forced to take into account the action of the whole automorphism group of the manifold. This is not very pleasant and one would like to have to deal with a smaller group of automorphisms, namely those isotopic to the identity.

Let us be more specific. Going back to Section 2.1 we know that an oriented standard spine determines an oriented 3-manifold up to oriented homeomorphism. So if we agree to fix a certain model of every manifold once and forever, then a spine P defines a unique manifold M, but the embedding $i : P \to M$ is determined only up to automorphisms of M. This suggests the following:

Definition 4.1.12. We call *embedded spine* a pair $(P, \overline{\imath})$, where P is an oriented standard spine and $\overline{\imath}$ is an equivalence class of embeddings of P into the corresponding oriented model manifold M, up to automorphisms of M isotopic to the identity.

The conclusion to which we are lead is the following: when the occasion demands, one might consider this notion of spine enhanced by the embedding, and refer to objects carried by the spine up to automorphisms isotopic to the identity. To restore the original viewpoint of spines it is then sufficient to quotient out by the whole automorphism group. The following is the embedded version of Theorem 4.1.9.

Corollary 4.1.13. *Let M be a fixed manifold. There exists a natural bijection between standard branched embedded spines of M and concave traversing flows on M up to isotopy with the same properties as in Theorem 4.1.9.*

4.2 Extending the flow to a closed manifold

We consider now the problem of extending the flow associated to a branched spine from a 3-manifold with boundary to a closed manifold which contains it. This is a very natural question which comes up, but there are also more specific motivations in connection with the notion of flow-spine which will be discussed in Sections 5.1 and 5.2. The proof of the following fact is easy.

Lemma 4.2.1. *Let N be a closed 3-manifold obtained by gluing two manifolds M and C along their homeomorphic boundaries. Consider a generic topological flow on N which induces a generic flow on M. Then also the flow induced on C is generic. Moreover points of $M \cap C = \partial M = \partial C$ which are convex (resp. concave) for M are concave (resp. convex) for C.*

Therefore we must investigate flows which have only convex points on the boundary (called *convex flows*). Another assumption we make is that the flow is a traversing one. The motivation for this is that we do not want anything essential to happen outside our manifold with boundary (for other motivations, see again Section 5.2).

Proposition 4.2.2. *Let C be a connected compact 3-manifold with a convex traversing flow. Then either C is a product $\Sigma \times [0,1]$ with Σ a closed surface and flow parallel to $[0,1]$ or it is a handlebody. In the latter case, if g is the genus of ∂C, there are exactly $1 + [g/2]$ inequivalent possibilities for the flow. In particular if ∂C is the 2-sphere or the torus then necessarily C is the 3-ball or the solid torus, with flow described in Figure 4.3.*

Proof of 4.2.2. Since there are no concave points, we can define a map $W \to B$ which assigns to a point of the white region W the only point of the black region B which belongs to the same orbit, and this map is of course a homeomorphism.

Fix a component W_0 of the white region. If W_0 has no boundary then of course the union of the orbits starting at W_0 is both open and closed in C, whence $W_0 = W$ and $C \cong W \times [0,1]$ with the prescribed flow.

If W_0 does have a boundary, recalling that ∂W_0 consists of convex points, we see that the union of ∂W_0 with all the orbits starting at W_0 is both open and closed in C. Therefore $W_0 = W$ and $C \cong \overline{W} \times [0,1]/\sim$, with $(x,t) \sim (x',t')$ if $x = x' \in \partial W$; the flow is the projection of $\{\{x\} \times [0,1] : x \in W\}$. In particular, C is homeomorphic to $\overline{W} \times [0,1]$ so it is a handlebody. Of course any connected surface W with boundary determines a unique C and the flow up to equivalence, and equivalent flows will have the same W. If W has genus h and n boundary components then ∂C has genus $g = 2h + n - 1$, so we have $1 + [g/2]$ possible different W's for a same g. $\boxed{4.2.2}$

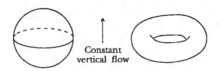

Figure 4.3: Convex flows on the ball and the solid torus

Remark 4.2.3. Consider a manifold M and the flow induced on it by an oriented-branched embedded spine as in Corollary 4.1.13. Assume it is possible to glue a manifold C to ∂M and extend the flow so that in C all the orbits are closed intervals. (For instance, ∂M could be S^2 with bicoloration consisting of one black disc, one white disc and one separating loop —this is the relevant case for flow-spines, see Section 5.1.) Then we have a finite number of choices for C and the flow on it up to homeomorphism (only one choice for $\partial M = S^2$). On the contrary, the flow on the resulting closed manifold is determined only in a much looser way. The reason is that even a small change in the gluing function might produce dramatic changes in the dynamics. Since the mapping class group of S^2 is trivial, *in the case $\partial M = S^2$, the flow happens to be determined up to homotopy* (which by definition means homotopy of the generating vector field in some smooth model). In general the equivalence is even looser, because there could be non-trivial elements of the mapping class group compatible with the bicoloration.

The next result deals with punctured closed manifolds, and should be compared with Proposition 4.2.2. Note that we are dropping the requirement that the flow should be traversing also on the "outside".

Proposition 4.2.4. *Let M be a closed oriented manifold, let N be M with one puncture and consider a branched standard spine P of N. Then the flow on N carried by P extends to M.*

Proof of 4.2.4. Let us consider the bicoloration (B, W, γ) induced on $S^2 = \partial N$. We know that $\chi(W) = \chi(B)$, so $\chi(B) = 1$. Moreover, a vector field on $S^2 = \partial B^3 \subset \mathbb{R}^3$ is a function $v : S^2 \to S^2$, which extends to B^3 if and only if it is null-homotopic, i.e. if it has degree 0.

In our situation, up to homotopy we can assume that γ is made of actual circles and that v is the identity in W and minus the identity in B, except in small neighbourhoods of the circles. Moreover we can imagine that all the circles are very close to the north pole, with the rest of the sphere white, so we can use planar pictures. In Figure 4.4 we show that a separating circle with black inside contributes with -1 to the degree, and a circle with white inside with $+1$. Let us think of a dynamic construction of the bicoloration starting with everything white, i.e. degree $d = 1$ and $\chi(B) = 0$. If we add a circle in the white or black region we have respectively:

$$\Delta d = -1, \Delta\chi(B) = 1, \qquad\qquad \Delta d = +1, \Delta\chi(B) = -1.$$

Since in the end $\chi(B) = 1$ we must have $\Delta\chi(B) = 1$, so $\Delta d = -1$ and $d = 0$, whence the conclusion. $\boxed{4.2.4}$

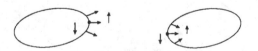

Figure 4.4: Contribution of separating circles

Using the Pontrjagin construction and the fact that $\pi_3(S^2) = \mathbb{Z}$, for any pattern on S^2 we can find fields on B^3 which match the pattern and are not homotopic relative to S^2. This implies that the extension constructed in this proposition is not unique up to homotopy.

Example 4.2.5. Let us consider in S^2 the bicoloration in which white and black region are both homeomorphic to the union of a disc and an annulus. An explicit extension in this case is obtained via the Hopf fibration $S^3 \to S^2$, because on a meridinal copy of $S^2 \subset S^3$ the induced pattern is the desired one.

4.3 Flow-preserving calculus: definitions and statements

In this section and two following ones we define a natural equivalence relation on flows associated to branched spines, and translate it into a calculus for normal o-graphs. This in particular yields a topological interpretation for some of the moves treated in Section 3.5. This is done in terms of the associated flow, introduced in the previous section. Let us recall that Theorem 4.1.9 establishes a correspondence between branched standard spines of M and concave traversing flows up to homeomorphism on M with certain additional properties.

Definition 4.3.1. We will consider on the set of concave traversing flows on a 3-manifold with boundary the relation of *homotopy through concave traversing flows*. To be precise: let $\{A_i^t\}$, $t = 0, 1$, be the flows; then we require that for some smoothing of the manifold there exists a smooth family $\{X_t : 0 \le t \le 1\}$ of smooth generic vector fields such that $\{A_i^t\}$ is the set of orbits of X_t for $t = 0, 1$, and the orbits of X_t are closed intervals for all t's.

Definition 4.3.2. We will call *standard sliding moves* on normal o-graphs the branched MP-moves of Figure 1.5, the move of Figure 1.4 (both in Chapter 1) and the inverses of them.

The motivation of this terminology will be evident in the sequel. We are now ready to state the main results of this chapter.

Theorem 4.3.3. *Two normal o-graphs define flows which are homotopic through concave traversing flows if and only if they are obtained from each other via a finite sequence of standard sliding moves.*

Corollary 4.3.4. *There exists a natural bijection between normal o-graphs up to standard sliding moves and concave traversing flows on 3-manifolds with boundary up to homeomorphism of the manifold and homotopy of the flow through concave traversing flows.*

The previous results can be stated in the refined category of embedded spines (see Definition 4.1.12) and proved by the very same arguments. To be precise one says that a normal o-graph is *embedded* if the corresponding spine is embedded up to isotopy in a fixed model of the corresponding manifold. The key remark is that the moves for standard spines and o-graphs naturally apply also to embedded spines and o-graphs (and will be called *embedded moves* in this context). Then the result is:

Corollary 4.3.5. *Let M be a fixed 3-manifold with boundary. There exists a natural bijection between embedded normal o-graphs of M up to embedded standard sliding moves and concave traversing flows on M up to homotopy through concave traversing flows.*

Before proceeding to the proofs we have a remark concerning the statement of Theorem 4.3.3 and some preliminary considerations in connection with Definition 4.3.1.

Remark 4.3.6. As a result of the application of the inverse of the move of Figure 1.4, one could end up with a single oriented loop without vertices, which is forbidden according to the original definition of o-graph. Note however that a single oriented loop, viewed as a *normal* o-graph, unambiguously defines a branched simple spine of a thrice punctured sphere. So to be completely formal in the statement of Theorem 4.3.3 we should either stipulate that the a move cannot be applied if the result has no vertices, or modify the definition to accept an oriented loop as a normal o-graph.

Remark 4.3.7. On a connected manifold, a generic flow such that all the orbits which touch the boundary are closed intervals is a concave traversing flow.

Lemma 4.3.8. *The contour curves of two flows homotopic through concave traversing flows are related by an isotopy of the boundary.*

Proof of 4.3.8. From Lemma 4.1.1 we know that the local model near a contour curve is :
$$M = \{(x, y, z) \in \mathbb{R}^3 : \ -\varepsilon < x, y < \varepsilon, f(x) \le z < \delta\}, \quad X = \partial/\partial x$$
with $\varepsilon, \delta > 0$, $f(0) = f'(0) = 0$ and f concave in $(-\varepsilon, \varepsilon)$. If $\tilde{X} = a \cdot (\partial/\partial x) + b \cdot (\partial/\partial y) + c \cdot (\partial/\partial z)$ is close to X, i.e. $a - 1, b, c$ are small, then the contour curve with respect to $\tilde{X}$ is given by
$$\{(x, y, f(x)) : \ f'(x) = c/a\}$$
which is isotopic to the original one. |4.3.8|

The converse of this lemma is not quite true, as we will see at the end of this section.

Remark 4.3.9. By genericity, given a homotopy through concave traversing flows, we can assume that there exist finitely many times $\{t_i\}$ such that:

1. At any time $t \notin \{t_i\}$, the interior of each orbit intersects the boundary in at most 2 points, and the contour curve is transversal to itself across any orbit segment (according to Definition 4.1.7);

2. At time t_i there is exactly one orbit which fails to satisfy the properties just stated. The interior of such an orbit meets the boundary in either 2 or 3 points, and there exists exactly 1 orbit segment across which the contour curve is not transversal to itself.

This fact will be used and explained by figures later in this chapter.

Now we turn our attention to the moves of Definition 4.3.2

Figure 4.5: A "rigid" normal o-graph

Remark 4.3.10. In a non-branched context the move of Figure 1.4 is generated by the MP-move, provided there are at least 2 vertices (this follows indirectly from the results of [45]). The same assertion is false in a branched context, at least in the framework of standard sliding moves, as can be seen from Figure 4.5. In fact, one sees that no sliding MP-move can be applied to the normal o-graph in this figure.

Remark 4.3.11. The move of Figure 1.4, even if actually independent of the other standard sliding moves, has the important property of being of a *strictly local nature*. This is a remarkable achievement of the choice of sticking to a standard context. In fact we will see that if one accepts also simple spines (as done for instance in [26]) non-local moves necessarily come into play.

Remark 4.3.12. The sum of the indices of the vertices is unchanged under the move of Figure 1.4. In Remark 3.5.4 this sum was noted to change by ± 1 under the moves of Figure 1.5. So the difference in the sum of indices of vertices gives a lower bound for the number of standard sliding moves which relate two normal o-graphs.

For the purpose of proving Theorem 4.3.3 we will need to extend our results and constructions to the case of *simple*, rather than standard, spines, even if the final result itself refers to standard spines only. The main reason why we prefer to deal with standard spines is that they are obviously determined by a neighbourhood of their singular set, and this fact allows to define the graphic presentation via o-graphs (or normal o-graphs if there is a branching). We will see in Section 5.1 that, even in the restricted setting of flow-spines, a simple branched spine is *not* determined by a neighbourhood of its singular set. Let us recall that a simple polyhedron is the same as a standard one, with the cellularity condition dropped.

We close this section with a note on the relation of homotopy through concave traversing flows. Using Lemma 4.3.8 and a collarization of the boundary we see that such a homotopy can be supposed to be constant on the boundary (or also a neighbourhood of the boundary). However homotopies constant on the boundary need not be through traversing flows:

Proposition 4.3.13. *Every concave traversing flow on a manifold with connected boundary is homotopic, via a homotopy fixed on the boundary, to non-traversing flows.*

Proof of 4.3.13. Assume that the flow defines a branched spine. Pick a generic point p_1 of the contour curve, so that the orbit starting from it intersects the boundary in a point q_1 of the black region B. Consider an arc α_1 properly embedded in B which connects q_1 to a generic contour point p_2 (here we are using connectedness of the boundary). Similarly proceed from p_2. Eventually a point q_n will be reached such that some p_i for $i \leq n$ is on the boundary of the component of B which contains q_n. Up to getting rid of some initial segment assume $i = 1$ and consider an arc α_n which joins q_n to p_1 (note that it might be $n = 1$; this is necessarily the case for instance if B is connected).

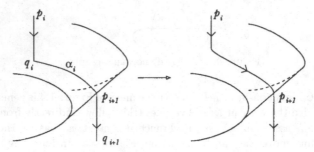

Figure 4.6: How to build a closed orbit

Then homotope the flow only in a small neighbourhood of the α_i's, as suggested by Figure 4.6. The resulting flow has a closed orbit which touches the boundary in finitely many concave points. This already is not a traversing flow, but one could also further homotope to get closed orbits which do not touch the boundary at all. $\boxed{4.3.13}$

4.4 Branched simple spines

The notions and results of this section are technical steps toward the proof of Theorem 4.3.3, but will also be needed later in Section 5.1.

Definition 4.4.1. We will say that a simple polyhedron is *oriented* if the smooth components are orientable and if a screw-orientation is given along the singular set (Definition 2.1.1).

Lemma 4.4.2. *An oriented simple polyhedron is the spine of a unique well-defined oriented manifold up to oriented homeomorphism.*

Proof of 4.4.2. Let P be oriented in the above sense. Let U be a closed regular neighbourhood of the singular set of P and let V_i's be the components of the closure of $P \setminus U$. Just as in the standard case we can canonically and properly embed U in an oriented handlebody H (the only difference is that now H may be disconnected). On ∂H we also have the boundary circles of the V_i's.

Let us choose on $V_i \times [-1, 1]$ any orientation. If γ is a component of ∂V_i and γ' is its image on ∂H, we have an essentially unique way to identify $\gamma \times [-1, 1]$ to a tubular neighbourhood of γ' in H so that $\gamma \sim \gamma \times \{0\} \sim \gamma'$ is the given identification, and the orientations inherited from H and from $V_i \times [-1, 1]$ are opposite to each other. Doing this for all the γ's we attach $V_i \times [-1, 1]$ to H, and the result is independent of the orientation initially fixed on $V_i \times [-1, 1]$ because this space admits an orientation-reversing automorphism. The attachment of all the V_i's produces the desired manifold, with orientation which extends the orientation of H.

Uniqueness follows from the fact that a $[-1, 1]$-bundle on an oriented surface (with boundary) whose total space is oriented is necessarily a product bundle. $\boxed{4.4.2}$

Remark 4.4.3. As opposed to the standard case [6], a simple polyhedron P which is the spine of an oriented 3-manifold may not be orientable in the above sense. In fact, any such P does have a screw orientation along the singular set, but the smooth components may not be orientable. As an example, the total space of the orientable $[-1, 1]$-bundle over $\mathbb{P}^2$ of course contains $\mathbb{P}^2$ as a simple spine.

The next definition paraphrases Corollary 3.1.7 and explains the reason why we have restricted to oriented spines in the above-defined sense.

Definition 4.4.4. An *oriented branching* on an oriented simple spine is an orientation for each of the smooth components such that along the singular set it never happens that the same orientation is induced three times.

The following fact is proved just as in the standard case.

Theorem 4.4.5. *Every oriented-branched simple spine of a 3-manifold M determines a concave traversing flow up to homeomorphism on M which satisfies properties 1-2 of Theorem 4.1.9, and conversely.*

We are now in a position to make the first step toward the proof of Theorem 4.3.3; namely we will now find the right set of moves for simple branched spines. In the next section we will check that in a standard context the standard sliding moves are sufficient. Let us first remark that the branched MP-moves of Figure 1.5 are of course well-defined also for simple branched spines.

Definition 4.4.6. We introduce for simple branched spines more moves, which we call *pure-sliding* moves, as shown in Figure 4.7. We will call *simple sliding moves* the moves of Figure 1.5, those of Figure 4.7, and inverses of them.

Some comments are in order about the definition of pure-sliding moves. First of all, the dotted arc is not actually part of the spine, it should be viewed as an instruction on how to perform the sliding. For the first two moves the arc carries all the information, while in the third move there is also a direction which tells which singular edge goes *over* the other one (the notion of 'over' involves the orientation of the smooth components and the screw-orientation along the singular set: the former has been indicated in the figure, the latter is the left-handed screw in $\mathbb{R}^3$).

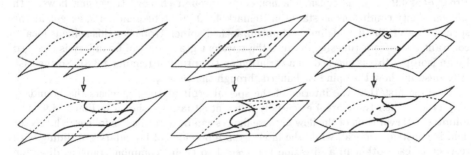

Figure 4.7: Moves on simple branched spines

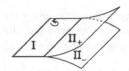

Figure 4.8: Types of edge for a disc

In Figure 4.8 we introduce some technical terminology, needed in the sequel in connection with the pure-sliding moves. If P is a simple branched spine, Σ is a smooth component (region) of P and e is an edge of Σ, we say that e is of *type* I, II$_+$ or II$_-$ for Σ according to the cases shown in the figure (e is the only edge shown, Σ is one of the three regions and on Σ we have written the type of e for Σ).

Remark 4.4.7. Here, and always in the rest of this section, we are deliberately ignoring the fact that a region could be multiply incident to an edge. To be completely formal one should introduce an abstract model of a region, where the edges are all distinct, and refer to a copy of the edge in the abstract model. This is a technical point which would obscure the ideas, so we leave it to the reader. Also, we keep calling edge of P a connected component of $S(P) \setminus V(P)$, even if in the simple case this may be a circle.

Using the notion of type, we see that a pure-sliding move is perfectly described by:
either: a dotted arc which joins an edge of type I with an edge of type II;
or: a dotted arc which joins an edge of type II$_+$ with an edge of type II$_-$;
or: a directed dotted arc which joins two edges of type I.

Remark 4.4.8. In general a pure-sliding move is not defined by only one (possibly directed) dotted arc. For instance, consider a vertex v of a branched spine and note that there are exactly two germs of discs D_1 and D_2 at v, opposite to each other, which have exactly one germ of edge of type I at v. Consider in D_i a dotted arc α_i, contained in a small neighbourhood of v, which joins the two edges incident to v. Then the pure-slidings along α_1 and α_2 have the same effect.

Proposition 4.4.9. *Two simple branched spines define flows which are homotopic through concave traversing flows if and only if they are related via a finite combination of simple sliding moves.*

Proof of 4.4.9. Let us consider a homotopy through concave traversing flows with the genericity conditions as stated in Remark 4.3.9. By Theorem 4.4.5, except at the special times t_i the flow defines a simple branched spine. Moreover this spine is locally constant away from times t_i, because the trivalent graph on the boundary is changed by an isotopy (this is checked as a variation on the proof of Lemma 4.3.8). So the task is to describe how the spine is changed through the time t_i.

Assume first that the interior of the special orbit at time t_i meets the boundary only twice. Then there are 4 essentially different situations, shown in Figure 4.9. The infinitesimal evolution of the flow in these figures can be described as follows: the vector field is constant vertical, while the intermediate portions of boundary are sliding with respect to each other in a direction transversal to their "common" tangent direction and to the field. If we construct the spines associated to the flows just before and just

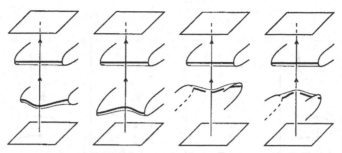

Figure 4.9: Special orbits

after time t_i, we very easily see that they always evolve via a sliding move. To be precise, the first two cases give the first move of Figure 4.7 (edge types I-II_ and I-II_+ respectively), the third case gives the second move (edge types II_+-II_-), and the fourth case gives the third move (edge types I-I).

We are left to deal with the case where the special orbit at time t_i meets another portion of boundary, necessarily in a contour curve. A complete 3-dimensional description of all possible patterns could be obtained by placing in each of the 4 situations of Figure 4.9 another fold transversely to the 2 already present. There are 3 different heights and 2 possible orientations in each case, so the total number sums up to 24. To avoid an excess of 3-dimensional figures we place ourselves on top of the flow and look down. A fold will be represented by an oriented curve, where the fold lies on the left of the curve. Note that these curves are just the contour lines, with coherent orientation, and our diagrams are exactly the associated o-graphs (the dashing and dotting should help figure out the actual 3-dimensional situation). The complete set of moves which arise is shown in the table of Figure 4.10 (we examine only one possible orientation for the new fold, the other one is similar). Each row corresponds to one of the slidings of Figure 4.9, whose pure form is shown in the first column. The second, third and fourth column correspond to the new fold being in a top, middle and bottom position.

We get 12 moves. Exactly 8 of them increase the number of vertices by 1, and it is readily seen that they are moves as in Figure 1.5, and conversely any such move is obtained (by considering also the other orientation for the new fold). The other 4 moves, which increase the number of vertices by 3, are easily shown to factor through simple sliding moves. An example is given in Figure 4.11, showing how to factor the move in position $(2, 4)$. $\boxed{4.4.9}$

Remark 4.4.10. From the point of view of simple branched spines, the whole set of moves of Figure 1.5 is redundant: using also the pure-sliding moves one easily shows that a few of the moves are sufficient to generate the others. We do not do this explicitly, because we most care of the standard case, where a similar fact fails to hold (see Remark 4.5.2).

Remark 4.4.11. The above theorem applies also to the case where the singular set is empty, so that the spine is just a closed surface. In this case we have that the spine is stable, and by Proposition 4.2.2 the associated manifold is just the product with the closed interval, with flow parallel to the interval.

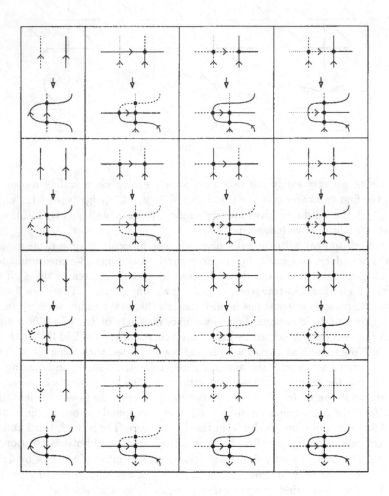

Figure 4.10: Special orbits interact with other orbits

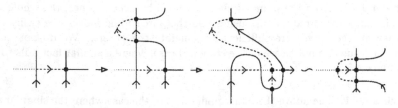

Figure 4.11: Factorization of a move

4.5 Restoring the standard setting

Having examined the simple case, we turn our attention to the standard one. We will have to show how to avoid pure-sliding moves, using the move of Figure 1.4.

Lemma 4.5.1. *If P is a standard branched spine, a pure-sliding move in P is obtained as a combination of standard sliding moves.*

Proof of 4.5.1. Let P be standard, Σ be a disc of P and a pure-sliding move on Σ be described by a dotted, possibly directed arc α. Assume first that α is incident to an edge e_0 of type I for Σ (if α is incident to two edges of type I, we take e_0 to be the one from which α starts). Let us number the other edges of Σ as $e_1, e_2, ..., e_p$ starting from e_0 according to the natural orientation of $\partial\Sigma$.

Let α_0 be the directed dotted arc in Σ which has both ends on e_0 and is directed so that on e_0 its target follows its source. For $i = 1, ..., p$ let α_i be the arc in Σ which joins e_0 and e_i, with direction from e_0 to e_i if also e_i is good. Finally, let α_{p+1} be the same as α_0 with reversed orientation. Of course our α is one of the α_i's.

We inductively prove that via standard sliding moves we can realize the pure-sliding along α_i. For $i = 0$ it is sufficient to perform the move of Figure 1.4 at e_0. The inductive step consists in applying two of the moves Figure 1.5 (a positive and a negative one) to make the second end of the dotted arc slip from e_i to e_{i+1}. One has to distinguish 9 cases according as these edges are of type I, II_+ or II_-. We do not do this explicitly in all cases, we only deal with the most delicate case of a transition from type I to type II or conversely. We actually prove a fact which is strictly speaking not necessary for our proof, but obviously true if an inductive process as that we are suggesting is to work. Recall that a pure-sliding move can be applied in one way only if one of the edges involved is of type II, and in two ways if both edges are of type I. This means that in our process of moving the dotted arc around the region, if we meet an edge of type II then we "lose memory" of the direction, if any. In Figure 4.12, using only normal o-graphs, we show that it is possible to "come out" of an edge of type II_- with any desired direction, the key point being that we meet a situation where the branched MP-move can be applied in two ways. The reader is invited to try and visualize what really is happening in this figure and complete the discussion of all the cases.

So the lemma is proved assuming α to touch an edge of type I. If this is not the case, let e (resp. e') be the edge of type II_- (resp. II_+) touched by α. If we apply

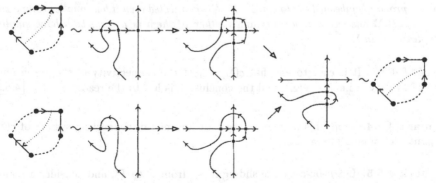

Figure 4.12. Construction of pure sliding moves: inductive step

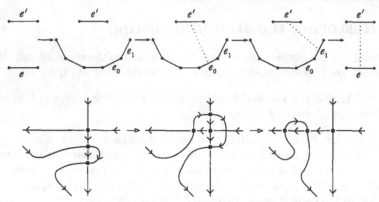

Figure 4.13: Construction of pure-sliding moves: the special case

the move of Figure 1.4 to e the effect from the point of view of Σ is to split e into 5 edges, of types respectively $II_-, II_-, II_-, I, II_-$, following the orientation (note that we are ignoring multiple adjacencies, as stated in Remark 4.4.7). We call e_0 and e_1 the last 2 edges of these 5. We know that via standard sliding moves we can realize the pure-sliding along the arc which joins e_0 and e'. Then we make the first end of this arc slip from e_0 to e_1, and we apply the inverse of the move of Figure 1.4 to get rid of the extra edges. Figure 4.13 illustrates this proof and shows that the slipping from e_0 to e_1 is possible.   4.5.1

Remark 4.5.2. In the above proof the whole set of moves of Figure 1.5 has been used to generate all pure-sliding moves. Therefore in the standard case one cannot play with pure-sliding moves to get rid of some MP-moves, so there is no obvious reduction of the number of moves (compare with Remark 4.4.10).

Before the final step of the proof of Theorem 4.3.3, we need an elementary fact. Let us call *complexity* of a connected surface Σ with boundary the minimal number $c(\Sigma)$ of disjoint properly embedded arcs cutting along which the surface becomes a disc.

Lemma 4.5.3. *Let Σ be a connected surface with non-empty boundary, and cut Σ along a properly embedded arc to get Σ'. If Σ' is connected then it has smaller complexity than Σ. If Σ has two components and neither of them is a disc, both have smaller complexity than Σ.*

Proof of 4.5.3. It is easy to see that $c(\Sigma)$ is just the connectivity of the graph onto which Σ retracts, i.e. $1 - \chi(\Sigma)$, and the conclusion is left to the reader.   4.5.3

Remark 4.5.4. From this lemma we see that any non-disconnecting system of $c(\Sigma)$ disjoint arcs cuts Σ into a disc.

Remark 4.5.5. Let μ be a simple sliding move from P to P', and consider a region R in P.

1. Assume μ is not a pure-sliding move along an arc in R; then it may happen that a small portion of R is cut out, or that R is connected to another region (in case μ^{-1} is a pure-sliding move), but anyway "most of R survives", and this allows to single out a region R' in P' of which R can be seen a subset of. If under μ^{-1} the same process leads to considering R' as a subset of R, then R and R' are homeomorphic, and the correspondence between them is natural.

2. This correspondence can be established also in case R has positive complexity and μ is a pure-sliding along a boundary-parallel dotted arc in R: in this case R is cut into two components one of which is a disc, and then we take R' to be the other component. Similarly the correspondence exists if μ^{-1} is a boundary-parallel pure-sliding in R' which gives R as non-disc region.

If R has positive complexity, we will say that a move μ as in 1 or 2 *does not reduce the complexity of* R, and we will regard R' as a copy of R survived through the move.

The following fact completes the proof of Theorem 4.3.3.

Proposition 4.5.6. *If two branched standard spines are related by a sequence of simple sliding moves then they are also related by a sequence of standard sliding moves.*

Proof of 4.5.6. Let $P_0 \to P_1 \to \cdots \to P_l$ be the sequence which exists by assumption. Let c be the maximal complexity of the P_i's. If $c = 0$ then Lemma 4.5.1 implies the conclusion. Let $c \geq 1$; we prove that there exists another sequence relating the same spines in which the maximal complexity is less than c, which allows to conclude by induction.

Let k_i be the number of regions of complexity c in P_i, and let k be the maximum of the k_i's. Let h be the number of P_i's for which $k_i = k$. Let us pick one of the P_i's with $k_i = k$, and in it a region of complexity c. Let us follow this region in both directions along the sequence, in the sense of Remark 4.5.5, as long as the region survives. Since $c > 0$ is maximal and in P_0 and P_l all the regions are discs, we find a segment of our sequence of the form

$$T \xrightarrow{\mu_0} T_1 \xrightarrow{\mu_1} \cdots \xrightarrow{\mu_{m-1}} T_m \xrightarrow{\mu_m} T'$$

with the following properties:

1. For $i = 1, ..., m$ there is in T_i a preferred region R_i of complexity c; the regions of complexity c are at most k in all and no region has complexity greater than c;

2. For $i = 1, ..., m - 1$ the move μ_i does not reduce the complexity of R_i, and R_{i+1} corresponds to R_i under the move;

3. The moves μ_0^{-1} and μ_m are pure-sliding moves which do reduce the complexity of R_1 and R_m respectively.

We will now show that under these assumptions we can find another sequence

$$T \xrightarrow{\nu_0} Q_1 \xrightarrow{\nu_1} \cdots \xrightarrow{\nu_{n-1}} Q_n \xrightarrow{\nu_n} T'$$

such that in Q_i there are at most $k - 1$ regions of complexity c and no region of complexity greater than c.

From this fact the conclusion of the proof follows very easily, because the argument allows to reduce h if $h > 1$, or k if $h = 1$, including the case $k = 1$, which yields a sequence of maximal complexity less than c.

So we have to construct this sequence $\{Q_i\}$. In R_1 let us consider the (possibly directed) arc along which the sliding μ_0^{-1} is performed. If μ_1 takes place far from R_1, or more in general if it does not affect the dotted arc, we can find a corresponding dotted arc in R_2, and proceed. Let us assume first that along the whole sequence the moves μ_i do not affect the dotted arc we have. We show in this case how to conclude, and then we indicate how to correct the argument when interaction takes place.

Let Q_i be obtained from T_i by sliding along the dotted arc we have in R_i. Then we just have $Q_1 = T$, so we take ν_0 to be the identity. Moreover since μ_i does not affect the dotted arc, it induces a move ν_i from Q_i to Q_{i+1}. At the end of the sequence we have in $R_m \subset T_m$ two possibly directed dotted arcs, which we denote by α and β, sliding along which we get T' and Q_m respectively. If α and β have no intersection then the slidings commute, i.e. we have in T' an arc which corresponds to β and in Q_m an arc which corresponds to α, sliding along which we get the same spine:

$$T' \xrightarrow[\nu_{m+1}^{-1}]{} Q_{m+1}, \quad Q_m \xrightarrow[\nu_m]{} Q_{m+1}$$

(the notation is chosen so to get directly the desired sequence).

Now let α and β have a finite number p of transversal intersections. We show that also in this case the pure-slidings can be realized so that they essentially commute. This is the key remark: one can think of a pure-sliding move on a region Σ in "physical" terms as follows: most of the spine, including Σ, is kept steady, while one region slides *over* Σ until it meets the boundary, and a bit more. But the same move can be described also by a region which slides *under* Σ.

In our situation let us stipulate that the pure-sliding along α is physically realized as an over-sliding, and that along β as an under-sliding. Then in T' we still see β, now divided into $1 + 2p$ subarcs, and the "physical" sliding along β can still take place, because the extra objects we have added are over R, and the sliding takes place under R. Of course this is a composition of $1+2p$ pure-slidings, whose final outcome we call S. In a similar way in Q_m we see α, and sliding along it we still get S. This construction is illustrated by an example in Figure 4.14. Now Lemma 4.5.3 easily implies that along the sequences which lead from T' and from Q_m to S we always have at most $k - 1$ regions of complexity c, so using these sequences we can complete the sequence of Q_i's

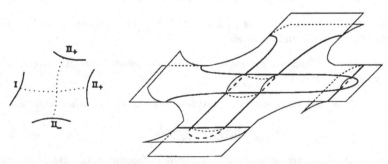

Figure 4.14: Commutation of physical slidings

we have been seeking.

We must now face the case when the moves do affect our dotted arc. By simplicity of notation we assume this interaction takes place for μ_1, and show how to correct the initial segment of the sequence $\{Q_i\}$ and which arc to consider in R_2. The same process can then by applied to carry on up to T_m, and the conclusion is the same as above. Recall that the dotted arc (call it α) in R_1 joins two edges. Then the only move which actually affects α is the collapse of one of these edges. So, assume μ_1 collapses an edge e_1 which contains an end of α, and let e_2 contain the other end. Let e_0 be an edge adjacent to e_1 and consider a dotted arc β parallel to α and disjoint from it, with its ends on e_0 and e_2. Assume that the pure-sliding along β can be performed. Then again we can exploit the commutation of slidings. We define:

$$Q_1 \xrightarrow{\nu_1} Q_2 \quad \text{along the arc which corresponds to } \beta;$$
$$T_1 \to Q_3 \quad \text{along } \beta \text{ (this move defines } Q_3, \text{ it is not in the sequence);}$$
$$Q_3 \xrightarrow{\nu_2^{-1}} Q_2 \quad \text{along the arc which corresponds to } \alpha;$$
$$T_2 \to Q_4 \quad \text{along } \beta \text{ (this move defines } Q_4, \text{ it is not in the sequence);}$$
$$Q_3 \xrightarrow{\nu_3} Q_4 \quad \text{move which corresponds to } \mu_1$$

(the last two definitions make sense now because μ_1 does not affect β). So we have defined the initial segment of the sequence, and we proceed from β in R_2.

Now, when does this construction fail to work? The only case is when e_2 and the two edges adjacent to e_1 are of the same type II_+ or II_-. Assume they are of type II_-. Since the pure-sliding along α is defined, we have that e_1 is of type I or II_+. But then, as shown in Figure 4.15, the situation is such that e_1 cannot be collapsed.

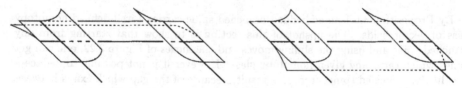

Figure 4.15: A non-collapsible edge

Above we have tacitly assumed that e_1 and e_2 are different edges. If they coincide then they must be of type I, and we leave to the reader to find a variation on the argument which allows to conclude also in this case. $\boxed{4.5.6}$

Proof of 4.3.4. Of course every concave traversing flow on a 3-manifold with boundary is homotopic through concave traversing flows to a flow which defines a simple branched spine. Then we just need to perform some pure-slidings so to turn this spine into a standard one. $\boxed{4.3.4}$

Proof of 4.3.5. We only need to remark that the relation of homotopy through concave traversing flows subsumes the relation of conjugation under automorphisms isotopic to the identity. $\boxed{4.3.5}$

4.6   The MP-move which changes the flow

In this section we will examine the homotopy class of the flow carried by a branched
spine, making use of the comparison class of vector fields on 3-manifolds which represents the primary obstruction for two fields to be homotopic, as defined below in
Section 6.1. Our result is that the MP-moves of Figure 3.22 (those which change the
bicoloration of the boundary), are sufficient to obtain all homotopy classes. However,
as we will point out, it is not possible to deduce from this fact the whole calculus for
branched spines.

In this section we will always refer to the notions of spine and move enhanced by
an embedding up to isotopy in a 3-manifold (Definition 4.1.12). The object we want to
study is the vector field which generates the flow associated to a branched spine. To
be formal about it, one needs to fix a smooth model of every manifold and restrict to
flows which are smooth in the fixed structure and compatible with the combinatorial
pattern given by the spine. We leave to the reader to fill in these details and prove
formally the following consequence of Corollary 4.1.13.

Proposition 4.6.1. *An embedded branched spine of a smooth 3-manifold M determines on M a nowhere-vanishing vector field, unique up to automorphisms of M isotopic
to the identity and rescaling, all the orbits of which are homeomorphic to closed intervals.*

Now we can rephrase Corollary 4.3.5 as follows:

Proposition 4.6.2. *Two embedded branched spines of a smooth 3-manifold are obtained from each other by a sequence of embedded standard sliding moves if and only
if the vector fields they define are homotopic through vector fields which have all the
orbits homeomorphic to closed intervals.*

By Proposition 4.6.1 an embedded branched spine carries a well-defined homotopy
class of vector fields. The upshot of this section is to show that starting from any
branched spine and using the sliding moves and the moves of Figure 3.22, one can get
spines which carry any given homotopy class. However it is not possible to deduce a
calculus for branched spines from this result, because of the gap which exists between
the notion of homotopy and the equivalence relation considered in Proposition 4.6.2.
Note that the gap is a priori not empty, because it contains the notion of homotopy
relative to the boundary (see also Proposition 4.3.13).

In the sequel we will freely use the results of Section 10.1 concerning the computation of homology, cohomology and Poincaré duality starting from any normal o-graph
of a 3-manifold with boundary. Moreover we will make use of the results and definitions of Section 6.1 (in particular the comparison class α of two vector fields, and
Proposition 6.1.8).

So, let us start analyzing the effect of the moves of Figure 3.22 on the homotopy
class of the vector field carried by a spine. Let us first remark that also these moves
have a physical interpretation as the sliding moves, and have a "pure version". This
pure version is shown in Figure 4.16: one should imagine that the horizontal plane
is steady and the two other regions are sliding over it until they collide: since it is

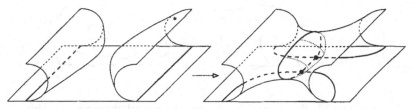

Figure 4.16: A region bumps into another one

impossible for them to slide over each other, what happens is that one bumps into the other. The actual moves of Figure 3.22 are obtained from the pure version by placing another region in a position where it does not affect the bumping (i.e. under the horizontal plane in Figure 4.16). In the sequel we will call *bumping moves* those of Figure 3.22.

Lemma 4.6.3. *Let two embedded branched spines P, P' of M be related by the move of Figure 4.16, and let $\overline{v}$, $\overline{v}'$ be the vector fields up to homotopy on M carried by these spines. Then $\alpha(\overline{v}', \overline{v}) = -[\hat{\Delta}]$ as an element of $H^2(P;\mathbb{Z}) \cong H^2(M;\mathbb{Z})$, where Δ is the region marked by the asterisk in Figure 4.16.*

Proof of 4.6.3. Essentially one only needs to use Lemma 6.1.3 and carefully look at the figure. Fix in Figure 4.16 coordinates (x, y, z) as shown in Figure 4.17, and consider

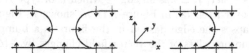

Figure 4.17: Cross-section of a bumping

the cross-section at $y = 0$ in the same figure. Here we have also chosen one of the two possible orientations for the discs and drawn the positive normal, which gives the vector field carried by the spine. We can assume that the trivializations of the tangent bundle given by the local embeddings in $\mathbb{R}^3$ are compatible with global trivializations, and that $(1, 0, 0)$ is a regular value. Then one sees that before the move the value is locally attained, with index $+1$, only in a point of Δ, and is not attained after the move, whence the conclusion. It is an interesting exercise to take some other regular value, e.g. $(-1, 0, 0)$ and show by means of the relations in $H^2(P;\mathbb{Z})$ that the result is the same. $\boxed{4.6.3}$

Theorem 4.6.4. *Given an embedded branched standard spine P of a manifold M and a homotopy class $\overline{v}$ of vector fields on M there exists a sequence of embedded standard sliding moves and embedded bumping moves which transforms P into an embedded branched spine which carries $\overline{v}$.*

Proof of 4.6.4. Let $\overline{w}$ be the class of the field carried by P, and let $\alpha(\overline{v}, \overline{w}) = [\sum_i n_i \hat{\Delta}_i]$ with $\{\Delta_i\}$ the smooth components of P. The idea is just to add (or subtract) the $\hat{\Delta}_i$'s one at a time, using a variation on Lemma 4.6.3, but one needs to be a bit careful because, as the spine changes, so does the presentation of the second cohomology group.

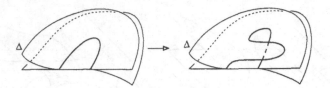

Figure 4.18: How to introduce a situation where bumping is possible

The key technical remark is the following: let $Q \to Q'$ be a pure-sliding move (in the direction which increases the number of vertices) or one of the moves of Figures 1.5 or 3.22 (also in the direction which increases the number of vertices). Then for every smooth component Δ of Q there exists a smooth component Δ' of Q' such that $[\hat{\Delta}] = [\hat{\Delta}']$ in $H^2(Q; \mathbb{Z}) = H^2(Q'; \mathbb{Z})$. The point is that all the regions survive (they may be cut into pieces by a pure-sliding, but they do not disappear), so the new presentation of H^2 is obtained by adding new generators, and relations which express these generators by means of the old ones.

Therefore, the theorem follows by induction from the following assertion: *given a region Δ of P and $\sigma = \pm 1$ there exists a sequence of pure-sliding moves and moves as in Figures 1.5 and 3.22 which leads from P to P', where, if $\overline{w}$ and $\overline{w}'$ are the homotopy classes of vector fields carried by P and P' respectively, we have $\alpha(\overline{w}', \overline{w}) = \sigma[\hat{\Delta}]$.*

To prove this assertion, we need to distinguish a few cases. We first deal with the case $\sigma = -1$, which is easier. If Δ has an edge e which is of type II for Δ, we consider the pure-sliding of e over itself (as in Figure 1.4), as shown in Figure 4.18. Then the MP-move which collapses the edge dashed in the figure is a bumping move, and by Lemma 4.6.3 the homotopy class of the vector field changes by $-[\hat{\Delta}]$ under the move (it does not matter which one, because Δ is actually bumping into itself).

If Δ is of type I for an edge e, and if Δ_1, Δ_2 are the other regions adjacent to e, we know that $[\hat{\Delta}] = [\hat{\Delta}_1] + [\hat{\Delta}_2]$ in H^2. Since we can realize the variation of α by $-[\hat{\Delta}_1]$ and by $-[\hat{\Delta}_2]$, we succeed also in this case.

Now let $\sigma = +1$. The idea is to produce a region Δ' with $[\hat{\Delta}'] = -[\hat{\Delta}]$, and then refer to the case $\sigma = -1$. To get such a Δ' we will produce a region which is zero in H^2, and then make Δ slide over it. A region which is zero in H^2 is the small region with 2 vertices in Figure 4.18, and this is what we will need. Let us see more in detail how to realize this strategy, because there is an interesting subtlety.

Again we can confine ourselves to the case where Δ has an edge e of type II (by simplicity, assume it is of type II$_+$). Call Δ_0 the region for which e is of type I. Assume first that Δ_0 has an edge e_0 which is of type II$_-$ for Δ_0. In this case our strategy can be put in practice directly, as shown in Figure 4.19. In the figure we are not showing the whole of the sliding of e_0 over itself, but only the part we need; Δ_1 is the disc which is zero in H^2 and Δ' is the disc we are seeking.

Now, assume Δ_0 has no edge of type II$_-$. Without spelling out the cases, we show in Figure 4.20 situations where an edge of type II$_-$ can be obtained by sliding. Inductively one easily sees that as soon as there are no monochromatic boundary components (black ones, in this case) it is always possible to achieve the condition by slidings. Therefore if the boundary is connected we always succeed. Contrarywise, if there is a monochromatic boundary component the reader can check that H^2 has a

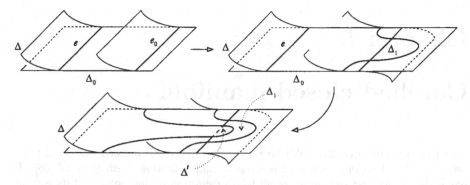

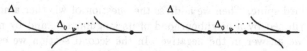

Figure 4.19: How to find a region opposite to a given one

Figure 4.20: How to obtain an edge of the needed type

direct summand $\mathbb{Z}$ generated by $[\hat{\Delta}]$, and that it is actually not possible to change α by $+[\hat{\Delta}]$ using sliding moves plus only one bumping move.

On the other hand, the argument can be easily adapted. If we first apply the procedure already described to change α by $-[\hat{\Delta}]$ then the bicoloration of the appropriate boundary component becomes non-trivial, and then we can apply twice the procedure which changes α of $+[\hat{\Delta}]$.

4.6.4

Chapter 5

Combed closed manifolds

This chapter is devoted to establishing a calculus for combed closed manifolds (Theorem 1.4.1) and to the relation of our results and definitions with those of [26]. In particular in the first section we recall Ishii's definition of flow-spine and its equivalence, *with the addition of the standardness condition*, with a suitably specialized notion of branched standard spine. Then we address the question of whether a non-standard flow-spine is determined by a neighbourhood of its singular set, and by means of an explicit example we answer in the negative. In the second section we establish the combinatorial realization of combed 3-manifolds as an application of the techniques developed in Chapter 4.

5.1 Simple vs. standard branched spines

In this section we provide an example of non-homeomorphic simple branched spines which have isomorphic neighbourhoods of the singular set. Simple branched spines have naturally come into play in Section 4.3 while analyzing the "flow-preserving" calculus for standard ones, and our example should be viewed as an evidence that one cannot base a combinatorial approach to the problem dropping the cellularity condition. Our example is actually in the restricted category of flow-spines, as defined in [24] and recalled below. For more motivation of this example in connection with Ishii's work [26], see the beginning of Section 4.1 and Remark 5.1.5 below. In the rest of this chapter and in the next one we will show that, adding the standardness condition, one can base on flow-spines calculi for oriented closed 3-manifolds and for vector fields up to homotopy on such manifolds.

We start with some preliminaries which establish relations between our notion of oriented branching and the notion of *flow-spine* of Ishii [24]. Let M be a smooth closed 3-manifold, ϕ a 1-parameter group of diffeomorphisms of M (generated by a nowhere-zero vector field X), and let Σ be a closed disc embedded in M. The triple (M, ϕ, Σ) is called *normal* if:

1. Σ (including the boundary points) is transversal to ϕ (namely: X is nowhere tangent to Σ).

2. Σ meets any half-orbit in positive and negative time (in formula: Given $x \in M$ there exist t_+, t_- with $t_- < 0 < t_+$ and $\phi_{t_\pm}(x) \in \Sigma$).

3. Any half-orbit in positive and negative time meets $\partial\Sigma$ at most 2 consecutive times (in formula: Given $x \in M$ and $t_0 < t_1 < t_2 < t_3$ such that $\phi_t(x)$ belongs to Σ for $t = t_0, t_3$, belongs to $\partial\Sigma$ for $t = t_1, t_2$, and does not belong to Σ for any other time between t_0 and t_3, then necessarily $\phi_{t_0}(x)$ and $\phi_{t_3}(x)$ belong to the interior of Σ).

4. $\partial\Sigma$ intersects itself transversely via the flow (in formula: If $x \in \partial\Sigma$, $t > 0$, $\phi_t(x) \in \partial\Sigma$ and $\phi_s(x) \notin \Sigma$ for $0 < s < t$, let v and w be vectors tangent to $\partial\Sigma$ at x and $\phi_t(x)$ respectively; then $X(\phi_t(x))$, $(d_x\phi_t)(v)$ and w should be linearly independent).

(We will also say that Σ is a *normal section* for (M, ϕ); in the language of [24], (ϕ, Σ) is called a normal pair on M.) The first statement of the next result is proved in Section 1 of [24] (we sketch the proof for completeness). The further statements establish the correspondence we need. We will say a bicoloration of S^2 is trivial if it consists of one black disc, one white disc and one separating loop.

Proposition 5.1.1. *1. Given a normal triple* (M, ϕ, Σ), *the set:*

$$P = \Sigma \cup \{\phi_t(x) : \ x \in \partial\Sigma, \ t > 0, \ \phi_s(x) \notin \Sigma \text{ for } 0 < s \leq t\}$$

is a simple spine of M *with one puncture.*

2. Under the same assumptions, if M *is oriented then* P *has a natural oriented branching, which induces the trivial bicoloration of the boundary sphere.*

3. Given an oriented-branched simple spine Q *of an oriented 3-manifold* N *bounded by a 2-sphere on which the induced bicoloration is trivial, there exists a normal triple* (M, ϕ, Σ) *such that* N *is homeomorphic to* M *with one puncture and* Q *is isomorphic, as an oriented-branched simple spine, to the spine* P *constructed in the previous points.*

Proof of 5.1.1. The fact that P is a simple spine is easy and graphically shown in Figure 5.1. Since P contains exactly the singularities of the function "first return time to Σ", it is not difficult to construct a regular neighbourhood U of P such that $(M \setminus U, \phi)$ is equivalent to $D^2 \times [0, 1]$ with the flow generated by the constant vector parallel to $[0, 1]$ (recall that Σ is a disc).

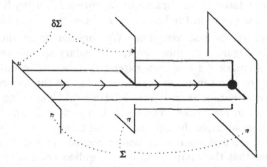

Figure 5.1: Proof that a flow-spine is a simple polyhedron

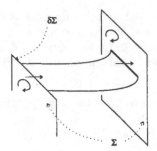

Figure 5.2: Oriented branching of a flow-spine

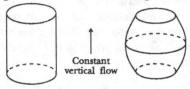

Figure 5.3: Flow on the complement of the spine, before and after smoothing

Let M be oriented. It follows from Figure 5.1 that every smooth component of P intersects Σ in a non-void open subset, and retracts onto this subset. Therefore, if we orient Σ so that the flow is positively transversal to it, we have a natural orientation for the smooth components of P. So P is oriented (recall from Remark 4.4.3 that the screw-orientation along the singularity comes with the orientation of M). To see that the orientations of the smooth components define a branching we remark that every component of the singular set minus the vertices contains arcs which lie in the interior of Σ. Therefore it is clear that two of the orientations along these arcs will cancel: see Figure 5.2. From the same figure (and taking into account also the behaviour at the vertices) one easily sees that the separating curve can be isotoped to $\partial\Sigma$, and therefore the bicoloration must be the trivial one.

It is worth mentioning another proof of the fact that the bicoloration is trivial, because it shades light on the inverse process (third assertion of the proposition). One can think of the "smoothing" of the spine of Figure 5.2 as obtained by slightly modifying the spine of Figure 5.1. Under this modification the complement of a regular neighbourhood evolves as in Figure 5.3. Since on the neighbourhood of P the flow is now the natural one induced by the branching (Figure 4.2), using Remark 4.1.11 and Figure 5.3 (right) we see that on the boundary we have the trivial bicoloration.

We are left to prove the last assertion. We consider on N the flow defined by the branching, as in Figure 4.2. Since on the boundary sphere we have the trivial bicoloration, using Remark 4.2.3 we can extend it to a flow ϕ on the manifold M obtained by capping off the sphere, in such a way that on the ball complementary to N we have the same situation as in Figure 5.3 (right). Now, using the flow, we cut the spine as described in Figure 5.4. The result is a surface Σ, and in order to verify that (M, ϕ, Σ) is a normal triple the only non-trivial thing to check is that Σ is a disc. But as in the proof of the first assertion, having removed the singularities of the return time function, we see that the complement of a neighbourhood of Q is homeomorphic to $\Sigma \times [0, 1]$, and on the other hand we already know that the complement is a ball.

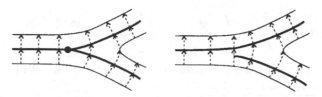

Figure 5.4: How to cut a spine to get a normal section

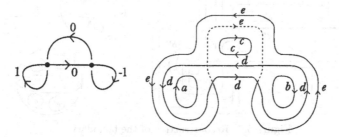

Figure 5.5: A normal o-graph and the data it encodes

From Figures 5.2 and 5.4 we immediately see that $P = Q$. 　　　　　5.1.1

Remark 5.1.2. An oriented-branched simple spine defines a normal o-graph, which carries a complete description of a neighbourhood of the singular set, including the oriented branching. One way to see this is to remark that to every such spine we can naturally associate an oriented-branched *standard* one: we just replace every smooth component by as many discs as its boundary circles, and orient in the obvious way. Using Remarks 3.2.2 and 3.3.3 and the previous proposition, we see that *an oriented-branched simple spine is a flow-spine if and only if it defines a manifold bounded by S^2 and the corresponding normal o-graph has only one circuit.*

Theorem 5.1.3. *There exist pairs of non-homeomorphic flow-spines having homeomorphic neighbourhoods of the singular set, where the homeomorphism preserves the branching (or, equivalently, the "flow-structure").*

Proof of 5.1.3. According to Remark 5.1.2, the strategy to obtain the counterexample must be as follows: start with a normal o-graph Γ with one circuit and find non-homeomorphic branched oriented *simple* spines P_1 and P_2 compatible with Γ such that the manifolds associated to P_1 and P_2 are both bounded by S^2. Note that the smooth components of the P_i's are necessarily planar (they are homeomorphic to proper subsets of S^2).

Let us consider the normal o-graph Γ of Figure 5.5–left (we have actually already met Γ in Figure 4.5, but now for better clarity we are specifying the colours of the edges). Denote by P the standard spine associated to Γ. A neighbourhood U of the singular set of P is described (with orientations given by the branching) in Figure 5.5–right; P is then obtained from U by attaching 5 discs with boundaries labeled by a, b, c, d, e. We now define oriented-branched simple spines P_1 and P_2 compatible with Γ. Both P_1 and P_2 differ from P for having an annulus $S^1 \times [0, 1]$ replacing 2 discs. The exact definition is given in Figure 5.6.

Figure 5.6: Smooth components to be glued to the singular set

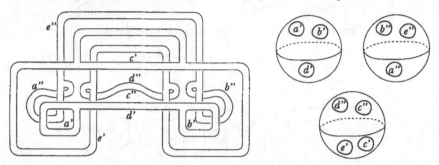

Figure 5.7: Reconstruction of the boundary

The verification that P_1 and P_2 are not homeomorphic, even disregarding the branching, is now easy, because in both the spines there is only one smooth component which is not a disc, and in P_1 this component is incident to 3 singular edges, while in P_2 it is incident to all 4 edges.

We are left to show that the manifolds defined by P_1 and P_2 are bounded by spheres. To this end we reconstruct the boundary of the manifold M defined by P, using Proposition 3.3.5, but also keeping track of the role of the various discs. This is done in Figure 5.7–left: remark that every disc D of P determines two discs D' and D'' on the boundary, where by notation D' is black (in other words, it projects onto D in an orientation-preserving way) and D'' is white. In the figure we have omitted the bicoloration, since we do not need it, but the reader could easily recover it using the labeling rule of discs. The result is the disjoint union of 3 spheres, in which the various discs play the role indicated in Figure 5.7–right.

Now in the boundary of the manifold defined by P_1 we will not have the discs bounded by a', a'', c', c'', but instead we will have annuli bounded by $(a', -c')$ and $(a'', -c'')$. Therefore the result is a sphere, as shown in Figure 5.8. For P_2 we must replace the discs bounded by d', d'', e', e'' by annuli bounded by $(d', -e')$ and $(d'', -e'')$. The result is again as in Figure 5.8, and the proof is complete. $\boxed{5.1.3}$

Remark 5.1.4. It is not difficult to see that the normal o-graph of Figure 5.5 represents a thrice punctured sphere S^3 (if you start with the obvious simple spine of this manifold, without vertices, and make it standard in the easiest way, you end up with the spine of this o-graph).

Figure 5.8: The boundary sphere

Remark 5.1.5. In the introduction to [26] the phrase "A flow-spine is completely determined by a data on the E-cycle", which forms the basis for the graphic calculus introduced in the paper, is contradicted by the example presented in this section (one just needs to remark that the "data on the E-cycle" are just a description of a neighbourhood of the singularity). For the proof Ishii addresses the reader to [24], where the result is found as a combination of Theorem 6.1 (and the construction which precedes it) with the phrase on page 523, lines 14-15. Theorem 6.1 takes as given a flow-spine, namely a normal pair on a 3-manifold, and shows how to explicitly and uniquely construct the map of which M is the mapping cylinder: in other words, this is a very particularized proof of the fact that a spine of the manifold has been constructed. However it may happen (and it does, according to our example) that non-isomorphic flow-spines have isomorphic neighbourhoods of the singular set. In the language of Ishii this means that the singularity data can be realized (as a flow-spine) in essentially different ways, whereas in [24], page 523, lines 14-15 he apparently asserts that a realization, if any, is unique.

Remark 5.1.6. It is to be noticed that our example contradicts what is stated in [26], because it shows that the singularity data do not determine the spine. However it does not contradict [24], page 523, lines 16-17 (namely, that the singularity data determine the manifold rather than the spine) because the two essentially different spines constructed in Theorem 5.1.3 both define B^3 (we leave this to the reader). We do not know if there exist counterexamples even to this weaker statement.

5.2 The combed calculus

This section is devoted to proving Theorem 1.4.1, namely that based on normal o-graphs there exists a calculus for homotopy classes of vector fields on closed oriented 3-manifolds. This is essentially a special case of Theorem 4.3.3, which takes into account Remark 4.2.3. The proof will use the notion of flow-spine introduced in Section 5.1, but the machinery developed in Chapter 4 to prove Theorem 4.3.3 (in particular Proposition 4.5.6) will also play an essential role.

Recall that spines a priori define manifolds only up to homeomorphism; therefore, if one works with plain spines, then he inevitably has to take into account the equivalence of combings which are conjugate under automorphisms. However, by exploiting the notion of embedded standard spine introduced earlier in this work, we can actually also treat more refined combings. We fix a representative for each equivalence class of closed oriented smooth 3-manifolds up to oriented diffeomorphism. We denote by $\mathcal{M}_{comb}^{emb}$ (resp. $\mathcal{M}_{comb}$) the set of pairs $(M, [v])$ with M a model manifold, and $[v]$ the equivalence class of a nowhere-zero vector field on M under the equivalence relation of homotopy through nowhere-zero vector fields (resp. homotopy and conjugation under smooth oriented automorphisms of M). Note that this definition of $\mathcal{M}_{comb}$ is compatible with that given in Chapter 1.

Denote by $\mathcal{G}_{comb}$ the set of normal o-graphs with the property that the boundary of the associated manifold is the 2-sphere with trivial bicoloration; we recall that 'trivial' means that there is one white disc and one black disc. We also note that the definition of $\mathcal{G}_{comb}$ is the same as given in Chapter 1, because:

- Axioms N1 and N2 imply that an object is in particular the normal o-graph of some branched standard spine P.

- Axiom C1 implies that the Euler characteristic of the boundary of the manifold N associated to P is 2.

- Axiom C2 implies that such a boundary is connected, so it is the 2-sphere, and therefore there exists a naturally associated closed manifold M.

- Axiom C3 implies that the bicoloration induced on ∂N by the field carried by P is trivial, so in particular the field naturally extends to M.

Note that by Theorem 4.3.3, Remark 4.3.6 and Proposition 3.5.1 the set $\mathcal{G}_{\text{comb}}$ is closed under standard sliding moves. We also recall that in Chapter 4 we have introduced the notion of embedded spine (Definition 4.1.12) and move (Section 4.3). In each of our model closed manifolds we make a puncture, so to get fixed models for all the manifolds bounded by S^2. We denote by $\mathcal{G}_{\text{comb}}^{\text{emb}}$ the set of pairs $(\Gamma, \bar{\iota})$, where Γ is an o-graph in $\mathcal{G}_{\text{comb}}$ and $\bar{\iota}$ is an embedding up to isotopy of the associated spine P in the corresponding model manifold (by definition $(P, \bar{\iota})$ is an embedded spine). We consider on $\mathcal{G}_{\text{comb}}^{\text{emb}}$ the set of embedded standard sliding moves. This implies Theorem 1.4.1:

Theorem 5.2.1. *There exists a natural bijection between $\mathcal{M}_{\text{comb}}^{\text{emb}}$ (resp. $\mathcal{M}_{\text{comb}}$) and the coset space of $\mathcal{G}_{\text{comb}}^{\text{emb}}$ (resp. $\mathcal{G}_{\text{comb}}$) under the equivalence relation generated by standard sliding moves (embedded moves in the case of $\mathcal{G}_{\text{comb}}^{\text{emb}}$).*

Recall that nowhere-zero vector fields up to homotopy correspond bijectively on a smooth 3-manifold to oriented plane fields up to homotopy, so our result could be equivalently stated in terms of oriented plane fields.

The proof of Theorem 5.2.1 follows quite closely the ideas of Ishii's work [26]. We note however that because of the counterexample shown in Section 5.1 the statements as they appear in the cited paper are incorrect: E-data do not uniquely describe spines, so no calculus for spines can be based on them. We also want to emphasize another important feature of o-graphs and normal o-graphs: such a graph automatically describes the spine of an oriented 3-manifold, and at least the first few properties of the manifold (such as the topology of the boundary, a presentation of the homology and cohomology groups, the bicoloration of the boundary in the case of normal o-graphs) are readily deduced. On the contrary in [26] it is declared that a priori one cannot even say when does an E-data define a manifold. Moreover the calculus for o-graphs, its version for normal o-graphs, and the special case of the latter which in view of Theorem 5.2.1 allows to describe homotopy classes of vector fields, are all generated by finitely many *local* moves, whereas if one accepts the point of view of simple spines, as done in [26], non-local moves should inevitably come into play.

To the end of proving Theorem 5.2.1 we will intensively use Proposition 5.1.1, of which in some sense the theorem is a refinement. We fix for the rest of the section one of the model closed manifolds M, and we denote by $N \subset M$ the manifold bounded by S^2 which we also use as model. We start with some technical preliminaries.

Lemma 5.2.2. *Let X be a vector field on M and $\gamma : [0, 1] \to M$ be a smooth arc such that $\dot{\gamma}(t)$ is not parallel to $X(\gamma(t))$ for $t = 0, 1$. Then there exists a smooth arc $\delta : [0, 1] \to M$ arbitrarily close to γ in the C^1-topology such that $\delta(t) = \gamma(t)$ and $\dot{\delta}(t) = \dot{\gamma}(t)$ for $t = 0, 1$, and $\dot{\delta}(t)$ is never parallel to $X(\delta(t))$ for $t \in [0, 1]$.*

Proof of 5.2.2. By compactness of $[0,1]$ we can assume $M = \mathbb{R}^3$. Since generically two vectors in 3-space are not parallel, for almost every $\alpha : [0,1] \to \mathbb{R}^3$ the arc $\gamma + \alpha$ fulfills the last condition, and the conclusion easily follows. $\boxed{5.2.2}$

Proposition 5.2.3. *Consider a flow ϕ on M. Then:*

1. *There exist normal sections of ϕ whose associated flow-spine is standard.*

2. *Given normal sections Σ_1 and Σ_2, there exist normal sections Δ_1 and Δ_2 such that*

$$\Delta_1 \cap \Sigma_1 = \Delta_2 \cap \Sigma_2 = \Delta_1 \cap \Delta_2 = \emptyset.$$

With regard to the second assertion of this proposition, it is claimed in [26] that given Σ_1 and Σ_2 one can find Δ disjoint from both. This is false: for a counterexample choose Σ_1 and Σ_2 such that $M \setminus (\Sigma_1 \cup \Sigma_2)$ is disconnected, and the closures of at least two connected components contain closed orbits.

Proof of 5.2.3. Let X be the vector field which generates ϕ, and consider on M a finite atlas made of cylinders $\{\phi_i : U_i \to D_\varepsilon^2 \times [0,1]\}_{i=0,1,...,k}$ such that under ϕ_i the field X corresponds to the constant field parallel to $[0,1]$ (here D_ε^2 is the closed disc of radius ε and we assume M is covered by the interiors of the U_i's). We further assume that in M the bases D_ε^2 are "very small" if compared to the heights $[0,1]$, and that the mutual intersections of the U_i's are "very small" if compared to the U_i's. Under these assumptions a generic choice of t_i's in $[0,1]$ produces a finite set $\{E_i = \phi_i^{-1}(D_\varepsilon^2 \times \{t_i\})\}_{i=0,1,...,k}$ of disjoint closed discs such that the union of their interiors intersects transversely all orbits in positive and negative time. For $i = 1, ..., k$ we select an arc $\alpha_i : [0,1] \to M$ so that:

1. $\alpha_i(0) \in \partial E_0$ and $\dot{\alpha}_i(0)$ is tangent to E_0 and points outwards;

2. $\alpha_i(1) \in \partial E_i$ and $\dot{\alpha}_i(1)$ is tangent to E_i and points inwards;

3. The images of all the α_i's are disjoint from each other and from the E_i's (except for the endpoints);

4. $\dot{\alpha}_i(t)$ is never parallel to $X(\alpha_i(t))$.

(The lemma is used for the last property.) Now it is not hard to find a thin strip s_i with core α_i to which X is never tangent, and in such a way that the union of the E_i's and s_i's is a closed disc Σ embedded in M (hint: introduce a metric and use the wedge). Of course Σ intersects transversely all the orbits in positive and negative time, and the other properties required to have a normal section are achieved by a generic small perturbation.

To conclude with the first assertion, we only need to note that via simple sliding moves we can turn any simple branched spine into a standard one, and recall Theorem 4.3.3 and Proposition 5.1.1.

The second assertion is established with the same method: one finds two systems of discs $\{E_i^1\}$ and $\{E_i^2\}$ as above, with

$$E_{i_1}^1 \cap \Sigma_1 = E_{i_2}^2 \cap \Sigma_1 = E_{i_1}^1 \cap E_{i_2}^2 = \emptyset \quad \forall i_1, i_2$$

and correspondingly systems of arcs α_i^1 and α_i^2 all disjoint from each other. We leave the details to the reader. $\boxed{5.2.3}$

Lemma 5.2.4. *Fix on M a Riemannian metric and consider two nowhere-zero vector fields X, Y on M. Let Σ be a normal section for the flow generated by X. Then there exists $\varepsilon > 0$ such that if $\|Y - X\|_{\infty,M} < \varepsilon$ then Σ is a normal section also for Y, and the corresponding flow-spines are isomorphic.*

Proof of 5.2.4. Let $T > 0$ be the maximal return time of Σ onto itself via X. Taking ε small enough we can assume that the Y-orbits of length $T + 1$ are arbitrarily close in C^1-sense to those of X. Now it is easily seen that the situations relevant for the construction of a flow-spine (Figures 5.1 and 5.2) are stable under very small C^1-perturbations of long enough orbits. $\boxed{5.2.4}$

Proof of 5.2.1. We prove the embedded version of the result, from which the non-embedded version immediately follows.

Since our model manifold N bounded by S^2 comes with an embedding in the corresponding closed manifold M, from Theorem 4.1.9 and Remark 4.2.3 we see that there exists a well-defined map from $\mathcal{G}_{\text{comb}}^{\text{emb}}$ to $\mathcal{M}_{\text{comb}}^{\text{emb}}$. Since all closed balls embedded in M are obtained from each other via automorphisms of M isotopic to the identity, Proposition 5.2.3 implies that this map $\mathcal{G}_{\text{comb}}^{\text{emb}} \to \mathcal{M}_{\text{comb}}^{\text{emb}}$ is surjective, and from Theorem 4.3.3 we deduce that there is an induced surjective map from $\mathcal{G}_{\text{comb}}^{\text{emb}}$ onto $\mathcal{M}_{\text{comb}}^{\text{emb}}/\sim$. We must show that this induced map is injective.

In force of Proposition 4.5.6 (and its embedded version) it is sufficient to prove the version of the theorem for simple spines, namely: *if two simple flow-spines define on M the same flow up to homotopy then they are obtained from each other via embedded simple sliding moves.* Another way to state the same assertion is the following: if ϕ_1 and ϕ_2 are homotopic flows on M and Σ_1, Σ_2 respectively are normal sections for them, the corresponding flow-spines are related by slidings. This is established in three steps.

First step: Case where Σ_1 and Σ_2 are disjoint sections of the same flow. With the same technique as in Proposition 5.2.3 we can join the sections by a strip to get a new section Σ. Then we can imagine a dynamic process in which we start with Σ_1, we expand it to get Σ and then we shrink it to get Σ_2. Along this process we can analyze the generic elementary catastrophes which cause essential changes in the spine corresponding to the section. The description is then exactly the same as in the proof of Proposition 4.4.9, where the role of the section is played by the black part of the boundary.

Second step: Case where Σ_1 and Σ_2 are any sections of the same flow. This is immediately deduced from the first step and the second assertion of Proposition 5.2.3.

Third step: Conclusion. According to the first two steps, a flow defines a unique flow-spine up to slidings. Since homotopy of flows means homotopy of the corresponding vector fields, Lemma 5.2.4 implies that the flow-spine up to slidings is locally constant along a homotopy, so it is constant. $\boxed{5.2.1}$

Chapter 6

More on combings, and the closed calculus

In this chapter we analyze in detail the set of combings on a closed oriented 3-manifold and provide a move iterating which any two combings are obtained from each other. The combinatorial translation of this move in terms of closed normal o-graphs, together with the same moves already considered for the combed case, will give the calculus for closed manifolds (Theorem 1.4.2). Some elementary results concerning combings are stated also for the case of manifolds with boundary (these have been used already in Section 4.6). We will also sketch the argument which gives the complete classification of combings in the closed case.

6.1 Comparison of vector fields up to homotopy

Let us fix a smooth connected 3-manifold M, possibly with boundary. We also fix a Riemannian structure on M and recall the obvious fact that a nowhere-vanishing vector field is homotopic to a unit vector field, and two unit vector fields are homotopic through unit vector fields if and only if they are homotopic through nowhere-vanishing vector fields. Therefore, if $\pi : UM \to M$ is the unit tangent bundle to M, our task is to study the collection $\Gamma(UM)$ of sections of UM, up to homotopy. The corresponding coset space will be denoted by $\mathcal{V}(M)$. It is also worth recalling that nowhere-zero vector fields correspond bijectively by orthogonality to oriented distributions of tangent 2-planes, so our results could be equivalently formulated in terms of these objects, which are of considerable interest for several reasons (e.g. contact structures are special distributions of 2-planes).

Since the tangent bundle to an oriented 3-dimensional manifold is trivial, there is a bijection between $\mathcal{V}(M)$ and the set $\pi(M; S^2)$ of homotopy classes of maps $M \to S^2$, but in general this bijection does depend on the trivialization.

Proposition 6.1.1. *Let $v_0, v_1 \in \Gamma(UM)$. Then:*

1. *Up to homotopy we can assume that $(-v_0)(M)$ and $v_1(M)$ are transverse submanifolds of UM;*

2. *$(-v_0)(M) \cap v_1(M)$ is a canonically oriented properly embedded 1-dimensional submanifold of UM, which properly embeds in M via π;*

3. *The class represented in $H_1(M, \partial M; \mathbb{Z}) \cong H^2(M; \mathbb{Z})$ by $\pi((-v_0)(M) \cap v_1(M))$ depends only on the homotopy class of v_0 and v_1.*

Definition 6.1.2. The class $\alpha([v_1], [v_0]) \in H^2(M; \mathbb{Z})$ which exists and is well-defined according to this result will be called the *comparison class* of the vector fields up to homotopy $[v_0]$ and $[v_1]$.

Proof of 6.1.1. Point 1 is obvious. In point 2, to define the orientation is a routine matter, and the other assertions are obvious. Point 3 is a general fact which in most abstract terms depends on the homotopy property of homological theories. $\boxed{6.1.1}$

If $b : UM \to M \times S^2$ is a trivialization of UM and $v \in \Gamma(UM)$ we denote by $v^{(b)} : M \to S^2$ the function such that $b(v(p)) = (p, v^{(b)}(p))$. Given a class $\bar{v}$ of vector fields up to homotopy, pick $v \in \bar{v}$. If $y \in S^2$ is a regular value of $v^{(b)}$, the preimage of y is a properly embedded oriented 1-dimensional submanifold of M, known to be independent up to oriented cobordism of y and of $v \in \bar{v}$. Therefore it defines a unique element of $H_1(M, \partial M; \mathbb{Z})$, and dually an element of $H^2(M; \mathbb{Z})$ which we will denote by $\beta^{(b)}(\bar{v})$.

Lemma 6.1.3. *For any trivialization b of UM and $v_0, v_1 \in \Gamma(UM)$ we have*

$$\alpha([v_1], [v_0]) = \beta^{(b)}([v_1]) - \beta^{(b)}([v_0]).$$

Proof of 6.1.3. Let $y \in S^2$ be a regular value for v_1 and $-v_0$. Up to homotopy we can assume that $\gamma_1 = v_1^{-1}(y)$ and $\gamma_0 = v_0^{-1}(-y)$ are disjoint, that $v_1 \equiv -y$ except in a small neighbourhood of γ_1 and that $v_0 \equiv y$ except in a small neighbourhood of γ_0. Then we have $v_1 = -v_0$ exactly on $\gamma_1 \cup \gamma_0$, and dealing with orientations one sees that $\alpha([v_1], [v_0]) = [\gamma_1] - [\gamma_0] \in H_1(M, \partial M; \mathbb{Z})$. $\boxed{6.1.3}$

If ξ is an oriented distribution of tangent 2-planes in M we will denote by $\mathcal{E}(\xi) \in H^2(M; \mathbb{Z})$ its Euler class.

Lemma 6.1.4. *For any trivialization b of UM and $v \in \Gamma(UM)$ we have:*

$$\mathcal{E}(v^\perp) = 2\beta^{(b)}([v]).$$

Proof of 6.1.4. We recall that the Euler class of ξ is the obstruction to the extension of a non-zero section of ξ from the 1-skeleton of a cellularization of M to the 2-skeleton. Let $y \in S^2$ be a regular value of both $v^{(b)}$ and $-v^{(b)}$. Let us view b as the restriction of a trivialization (called b again) of the tangent bundle TM. The section of $v^\perp$ given by

$$s : M \ni p \mapsto b^{-1}(p, y \wedge v^{(b)}(p)) \in TM$$

vanishes exactly at the points of the oriented properly embedded submanifold

$$Z = (v^{(b)})^{-1}(y) \cup (v^{(b)})^{-1}(-y)$$

which can be assumed not to meet the 1-skeleton of a cellularization and be transversal to the 2-cells. If p is a point of a 2-cell where s vanishes, an easy computation shows that the index of the zero is equal to the sign of the intersection of Z with the 2-cell. This allows to conclude at once. $\boxed{6.1.4}$

Corollary 6.1.5. *For any $v_0, v_1 \in \Gamma(UM)$ we have:*

$$2\alpha([v_1], [v_0]) = \mathcal{E}(v_1^\perp) - \mathcal{E}(v_0^\perp).$$

Proposition 6.1.6. *For $v \in \Gamma(UM)$ the class $\beta^{(b)}([v])$ in general does depend on the trivialization b of UM, but it does not if $H^2(M; \mathbb{Z})$ is free of 2-torsion.*

Proof of 6.1.6. The second assertion follows immediately from the previous results. For an example where $\beta^{(b)}$ depends on b (in some sense a prototype) consider as a manifold the real projective space $\mathbb{P}^3$. Note that $H^2(\mathbb{P}^3; \mathbb{Z}) \cong \mathbb{Z}_2$, and fix any trivialization $b : U\mathbb{P}^3 \to \mathbb{P}^3 \times S^2$. Recall that $\mathbb{P}^3$ is diffeomorphic to the special orthogonal group $SO(3)$, and fix a diffeomorphism ϕ. Consider a new trivialization c of $U\mathbb{P}^3$ defined by $c(b^{-1}(p, y)) = (p, \phi(p)(y))$. Let v be a vector field such that $v^{(b)}$ is constant, so that $\beta^{(b)}([v]) = 0$ in $H^2(\mathbb{P}^3; \mathbb{Z})$. On the contrary, $\beta^{(c)}([v])$ is dual to the loop in $SO(3)$ given by rotations on a certain plane, and hence it is non-zero in $H^2(\mathbb{P}^3; \mathbb{Z})$. $\boxed{6.1.6}$

Lemma 6.1.7. *Let $v, w \in \Gamma(UM)$. If $[v] = [w]$ then $\alpha([w], [v]) = 0$. If $\partial M \neq \emptyset$ the converse also holds.*

Proof of 6.1.7. The first assertion is obvious. Assume $\partial M \neq \emptyset$ and $\alpha([v], [w]) = 0$, and consider a branched standard spine P of M (any 2-dimensional spine would do the job, we use P to refer to notations we have already introduced and the homology computations which will be carried out in Section 10.1). Fix a trivialization of UM and denote by v and w again the resulting maps $M \to S^2$. Since P is a deformation retract of M it is sufficient to show that the restrictions of v and w to P (denoted by v and w again) are homotopic. Let $y \in S^2$ be a regular value for v and w. In particular the value y is not attained on $S(P)$, so we can assume v and w to have constant value $-y$ on a neighbourhood of $S(P)$. Let $\{\Delta_i\}$ be the discs of P. Then $v^{-1}(y) \cap \Delta_i$ and $w^{-1}(y) \cap \Delta_i$ consist of a finite set of points with index ± 1. Let n_i and m_i respectively denote the sums of these indices. The assumption $\alpha([w], [v]) = 0$ now reads

$$[n_1 \hat{\Delta}_1 + \cdots + n_k \hat{\Delta}_k] = [m_1 \hat{\Delta}_1 + \cdots + m_k \hat{\Delta}_k] \text{ in } H^2(P; \mathbb{Z}).$$

If $n_i = m_i$ then by the Hopf Theorem we have that v and w are homotopic on Δ_i with homotopy fixed on $\partial \Delta_i$. Figure 6.1 suggests how to translate a relation $[\hat{\Delta}] =$

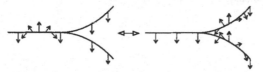

Figure 6.1: Translating a relation into a homotopy

$[\hat{\Delta}_+] + [\hat{\Delta}_-]$ in H^2 into a homotopy on w which changes the m_i's accordingly. Hence we can assume $m_i = n_i$ for all i, and then conclude. $\boxed{6.1.7}$

Proposition 6.1.8. *For every $\overline{v} \in \mathcal{V}(M)$ the map*

$$\mathcal{V}(M) \ni \overline{w} \mapsto \alpha(\overline{w}, \overline{v}) \in H^2(M; \mathbb{Z})$$

is surjective, and if $\partial M \neq \emptyset$ it is also injective.

Proof of 6.1.8. Surjectivity follows from the Pontrjagin construction [39] (see also the next section) and injectivity in case $\partial M \neq \emptyset$ follows from Lemma 6.1.7. $\boxed{6.1.8}$

6.2   Pontrjagin moves for vector fields, and complete classification

In this section we will prove that on a closed manifold any two vector fields up to homotopy are related by finitely many applications of a certain move, which essentially corresponds to the Pontrjagin construction. This will be used, in the next section, to construct a finite local calculus for closed oriented 3-manifolds based on closed normal o-graphs.

For the purpose of the application, we will need an intrinsically defined version of the move, i.e. one which is independent of a trivialization (however a trivialization will come in the proof, in a way we shall point out). If one accepts a trivialization from the beginning then one can prove the same result, along the same lines, using as a move exactly the Pontrjagin construction. Let us start with the definition of the move and the main statement.

In the rest of this section M is a connected, closed, oriented smooth 3-manifold. We will make use of the notions concerning vector fields on M introduced in the previous section. Let us just recall here that $\mathcal{V}(M)$ denotes the set of vector fields up to homotopy on M (when the occasion demands we will use only unit fields with respect to some fixed Riemannian structure).

Definition 6.2.1. Let $\bar{v} \in \mathcal{V}(M)$, let $v \in \bar{v}$ and let $j : D^2 \times S^1 \to M$ be an embedding such that $v \circ j = j_*(\partial/\partial\varphi)$, where φ is the coordinate of $S^1 = \mathbb{R}/2\pi\mathbb{Z}$. Define a new field v' which coincides with v outside the image of j and there is defined by

$$v'\big(j\big(\rho e^{i\vartheta}, e^{i\varphi}\big)\big) = j_*\big(-\cos(\pi\rho)\frac{\partial}{\partial\varphi} - \sin(\pi\rho)\frac{\partial}{\partial\rho}\big).$$

Then we will say that $[v'] \in \mathcal{V}(M)$ is obtained from $\bar{v}$ by a *Pontrjagin move*.

The qualitative effect of the Pontrjagin move is shown in Figure 6.2: on the left one sees a cross-section of the vector field and on the right the orbits in a set of the form $\{te^{i\vartheta} : -1 \le t \le 1\} \times \mathbb{R}$, where $\mathbb{R}$ is the universal cover of S^1 (by definition the field lifts to the universal cover).

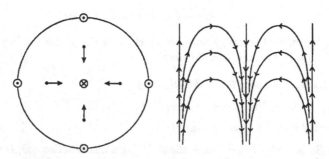

Figure 6.2: Field produced by the Pontrjagin move in a torus

Theorem 6.2.2. *Any two elements of $\mathcal{V}(M)$ are obtained from each other by a finite combination of Pontrjagin moves.*

This statement appears in [26], but we could not understand the proof outlined there. We also think that the method used below gives an interesting insight to the situation.

The structure of our proof is as follows. Given two fields up to homotopy, we first show that via Pontrjagin moves we can reduce to the case where the fields have zero comparison class in $H^2(M; \mathbb{Z})$. Then we prove that two such fields can be assumed to coincide on the complement of a ball, so that, in a certain trivialization, their difference can be interpreted as an element of $\pi(B^3, S^2; S^2, \{*\}) \cong \pi_3(S^2) \cong \mathbb{Z}$, and then the last step is to show that the Pontrjagin move is sufficient to generate the whole of $\pi_3(S^2)$.

In the sequel we will repeatedly use the notions and notations introduced in the previous section.

Lemma 6.2.3. *Given $\bar{v}, \bar{w} \in \mathcal{V}(M)$ there exists a finite sequence of Pontrjagin moves which leads from $\bar{w}$ to a $\bar{z} \in \mathcal{V}(M)$ such that $\alpha(\bar{z}, \bar{v}) = 0$ in $H^2(M; \mathbb{Z})$.*

Proof of 6.2.3. Let $\alpha(\bar{w}, v) = h \in H^2(M; \mathbb{Z})$. We must obtain a $\bar{z}$ such that $\alpha(\bar{w}, \bar{z}) = h$. Assume by simplicity that dually in $H_1(M; \mathbb{Z})$ the class h is represented by only one embedded loop $\gamma : S^1 \to M$. Fix a representative w of $\bar{w}$. Up to a small perturbation of γ we can assume that w is never parallel to $-\dot{\gamma}$. Then up to a homotopy on w we can assume that w equals $\dot{\gamma}$ on γ, and we can further homotope w so that on a tubular neighbourhood $D^2 \times S^1 \hookrightarrow M$ of γ it is the constant field parallel to the S^1 factor. Now let us apply the Pontrjagin move using the representative w and the tube thus obtained, and let us call z the result. Of course we have $w = -z$ exactly on γ, and dealing with orientations one easily sees that $\alpha(w, z) = h$ indeed. $\boxed{6.2.3}$

We have shown that via Pontrjagin moves we can remove the first obstruction to homotopy. We have now a second obstruction, to deal with which we fix a trivialization of the tangent bundle So now we consider the set $\pi(M; S^2)$ of maps $M \to S^2$ up to homotopy, and for every $f : M \to S^2$ we have a corresponding $\beta([f]) \in H^2(M; \mathbb{Z})$. According to Lemma 6.1.3 we are left to study the fibres of β. To this aim we consider an embedded closed ball B in M, we denote by N the closure of $M \setminus B$ and by $i : N \to M$ the inclusion. We will use below the induced homomorphism $i^* : H^2(M; \mathbb{Z}) \to H^2(N; \mathbb{Z})$. Using the Mayer-Vietoris exact sequence one sees quite easily that i^* is actually an isomorphism. We also fix a basepoint e_0 in S^2.

The following preliminary result follows from Proposition 6.1.8, together with an easy argument which we leave to the reader.

Lemma 6.2.4. *For every $h \in H^2(M; \mathbb{Z})$ there exists $f_h : M \to S^2$ such that $\beta([f_h]) = h$ and $f_h = e_0$ on B;*

Let us fix these f_h's arbitrarily.

Lemma 6.2.5. *If $\bar{f} \in \pi(M; S^2)$ and $\beta(\bar{f}) = h$ then there exists $f \in \bar{f}$ such that $f = f_h$ on N.*

Proof of 6.2.5. Let us consider the comparison classes α_M and α_N. From the definition it is quite obvious that for any $v, w : M \to S^2$, viewed as vector fields via the trivialization b, we have:

$$\alpha_N \left([w|_N], [v|_N] \right) = i^* \left(\alpha_M([w], [v]) \right).$$

So in our situation, using Lemma 6.1.7, given $f' \in \bar{f}$, we see that the restrictions to N of f' and f_h are homotopic. Let F be the homotopy, with $F(0, \cdot) = f'$ and $F(1, \cdot) = f_h$. We now define a map $G : [0, 1] \times M \to S^2$ by:

$$G(t, x) = \begin{cases} F(t, x) & \text{if } x \in N \\ F(2|x| - 2 + t, x/|x|) & \text{if } x \in B, \ |x| \geq 1 - t/2 \\ f'(x/(1 - t/2)) & \text{if } x \in B, \ |x| \leq 1 - t/2 \end{cases}$$

where by abuse of notation we are identifying B with B^3. Of course G is a homotopy between f' and a map f as required. $\boxed{6.2.5}$

We denote by $\pi(B, \partial B; S^2, e_0)$ the set of maps $B \to S^2$ which take the value e_0 on ∂B, up to homotopy relative to the boundary. Note that since $B/\partial B \cong S^3$ this set can be identified to $\pi_3(S^2)$, and hence to $\mathbb{Z}$.

For $h \in H^2(M; \mathbb{Z})$ and $g : B \to S^2$ with $g(\partial B) = e_0$, we now define $f_{(h,g)} : M \to S^2$ as the map which coincides with f_h on N and with g on B. The following result summarizes our construction. Its proof is straight-forward using the previous lemmas and the definitions.

Proposition 6.2.6. *The map*

$$\Psi : H^2(M; \mathbb{Z}) \times \pi(B, \partial B; S^2, e_0) \ni (h, [g]) \mapsto [f_{(h,g)}] \in \pi(M; S^2)$$

is well-defined and surjective.

We are now ready for the final step toward the main result of this section.

Proof of 6.2.2. Essentially what we are left to show is that via Pontrjagin moves we can reconstruct the whole of $\pi(B, \partial B; S^2, e_0)$, but we must be a bit careful because the move is defined intrinsically, while $\pi(B, \partial B; S^2, e_0)$ assumes a trivialization of the tangent bundle. However, since B is contractible, up to homotopy we can assume from the beginning that B is identified to B^3 and that on it the trivialization is given by the embedding in $\mathbb{R}^3$.

We now recall how $\pi(B^3, \partial B^3; S^2, e_0)$ is identified to $\mathbb{Z}$, paraphrasing one of the descriptions of $\pi_3(S^2)$. If $g : B^3 \to S^2$ is smooth and y is a regular value, $g^{-1}(y)$ is a link L in the interior of B^3 with a framing ν (a normal vector field to L), and the integer corresponding to $[g]$ is the linking number of L and $L + \varepsilon\nu$ for small ε.

We show that starting from the constant map e_0 via a Pontrjagin move we can obtain any integer n. Since at least once in this work we can carry out a proof with precise formulae without using figures, we do this in detail. By simplicity of notation replace the ball B^3 by another ball with big radius. Consider the following loops in $\mathbb{R}^3$:

$$\gamma_\pm : [0, 2\pi] \ni \varphi \mapsto 3(0, \cos\varphi, \pm\sin\varphi) \in \mathbb{R}^3.$$

Note that γ_+ (resp. γ_-) bounds a disc to which the direction e_0 is positively (resp. negatively) transversal, whence the notation. Let us parametrize a tubular neighbourhood of $\gamma_\pm$ as:

$$j_\pm : [0,2] \times [0, 2\pi] \times [0, 2\pi] \ni (\rho, \vartheta, \varphi)$$
$$\mapsto (3 + \rho \cos \vartheta)(0, \cos \varphi, \pm \sin \varphi) + (\rho \sin \vartheta, 0, 0) \in \mathbb{R}^3.$$

Now, by taking convex combinations in S^2 on the region $1 \leq \rho \leq 2$ we can homotope the constant field e_0 to fields $e_\pm^{(0)}$ with everywhere non-negative e_0-component and:

$$e_\pm^{(0)}(j_\pm(\rho, \vartheta, \varphi)) = (0, -\sin \varphi, \pm \cos \varphi) = \dot\gamma_\pm(\varphi)/3 \quad \text{for } \rho \leq 1.$$

We are now essentially in the situation in which the Pontrjagin move can be applied to the torus $\{\rho \leq 1\}$. In fact, the orbits have the required shape, and to strictly satisfy the conditions we only have to rescale the field. The effect of the move is to produce a new field $e_\pm^{(1)}$ which coincides with $e_\pm^{(0)}$ outside $\{\rho \leq 1\}$, and is given there by:

$$e_\pm^{(1)}(j_\pm(\rho, \vartheta, \varphi))$$
$$= -\cos(\pi \rho)(0, -\sin \varphi, \pm \cos \varphi) - \sin(\pi \rho)(\sin \vartheta, \cos \vartheta \cos \varphi, \pm \cos \vartheta \sin \varphi).$$

(Again, this is not quite the field one gets as a direct application of the definition, but the orbits are qualitatively those we want, so up to homotopy this is the right field.)

The value $(-1, 0, 0)$ is a regular one for $e_\pm^{(1)}$, and it is attained exactly on the curve

$$\delta_\pm : [0, 2\pi] \ni \varphi \mapsto j_\pm(1/2, \pi/2, \varphi) = (1/2, 3 \cos \varphi, \pm 3 \sin \varphi)$$

(note that δ_+ and δ_- have the same support). Now a direct computation shows that:

$$\left(d_{(1/2, \pi/2, \varphi)}(e_\pm^{(1)} \circ j_\pm) \right) \left(-\frac{\sin \varphi}{\pi} \frac{\partial}{\partial \rho} + \cos \varphi \frac{\partial}{\partial \vartheta} \right) \equiv (0, 1, 0)$$

and therefore the framing on $\delta_\pm$ is given by the following normal field:

$$\nu_\pm(\varphi) = -(1, 0, 0)(\sin \varphi)/\pi - (0, \cos \varphi, \pm \sin \varphi)(\cos \varphi)/2.$$

Then it is easy to see that

$$\text{Lk}(\delta_+, \delta_+ + \nu_+) = -1 \qquad \text{Lk}(\delta_-, \delta_- + \nu_-) = +1.$$

We can therefore conclude that starting from the constant field e_0 the element of $\pi(B, \partial B; S^2, e_0)$ which corresponds to an integer n can be realized by $|n|$ Pontrjagin moves on suitably chosen unknots. $\boxed{6.2.2}$

We conclude this section by coming back to Proposition 6.2.6. It follows from its statement that for for $h \in H^2(M; \mathbb{Z})$ the map:

$$\Psi_h : \mathbb{Z} \cong \pi(B, \partial B; S^2, e_0) \ni [g] \mapsto [f_{(h,g)}] \in \{\bar{f} \in \pi(M; S^2) : \beta(\bar{f}) = h\}$$

is surjective. So the question which naturally arises is whether this map is injective. To answer, we will outline a simple geometric argument suggested to us by Alexis Marin which leads to a complete description of $\pi(M, S^2)$.

So, consider maps $f_0, f_1 : M \to S^2$ such that $\beta(\overline{f}_0) = \beta(\overline{f}_1) = h$. We can consider a value $y \in S^2$ which is regular for both f_0 and f_1, so we get homologous oriented links $L_0 = f_0^{-1}(y)$ and $L_1 = f_1^{-1}(y)$. Let S be an oriented surface properly embedded in $M \times [0, 1]$ which realizes the homology between $L_0 \subset M \times \{0\}$ and $L_1 \subset M \times \{1\}$. This means that

$$\partial S = S \pitchfork \partial(M \times [0, 1]) = L_0 \cup (-L_1).$$

Recall now that along L_0 and L_1 we have framings defined by normal fields ν_0 and ν_1. We can think of ν_0 and ν_1 as tangent fields to $M \times [0, 1]$ along ∂S; note also that they are transversal to S. It is not hard to see that we can extend ν_0 and ν_1 to a tangent field ν to $M \times [0, 1]$ along S which is transversal to S.

We can use such a field ν to perturb S and hence compute the corresponding self-intersection number of S, denoted by $n = n(S, \nu) \in \mathbb{Z}$. To see how n depends on our choices, take another pair (S', ν') which satisfies the same properties as (S, ν), and denote by n' the corresponding integer. We can glue together along the boundary two copies of $M \times [0, 1]$ matching (S, ν) with (S', ν') along $L_0 \cup L_1$; this leads to a closed oriented surface $\Sigma \subset M \times S^1$ and a vector field η along Σ which is generically transverse to Σ. If we perturb Σ using η and compute the self-intersection, we get that

$$n - n' = [\Sigma] \cdot [\Sigma]$$

where $[\Sigma] \in H_2(M \times S^1; \mathbb{Z})$ and $[\Sigma] \cdot [\Sigma] \in \mathbb{Z} = H_4(M \times S^1; \mathbb{Z})$. Now using the Künneth formula in $M \times S^1$ we can write

$$[\Sigma] = [a \times S^1] + [b \times \{1\}]$$

where a and b are respectively a 1-cycle and a 2-cycle in M. Now note that

$$[\Sigma] \cdot [M \times \{1\}] = [L_0 \times \{1\}] = i_*(h)$$

where $h = [L_0] = [L_1]$ and $i : M \hookrightarrow M \times \{1\} \subset M \times S^1$ is the inclusion. Moreover:

$$\begin{aligned} [\Sigma] \cdot [M \times \{1\}] &= [a \times S^1] \cdot [M \times \{1\}] + [b \times \{1\}] \cdot [M \times \{1\}] \\ &= [a \times \{1\}] + 0 = i_*([a]) \end{aligned}$$

therefore $[a] = h$. Now

$$\begin{aligned} [\Sigma] \cdot [\Sigma] &= [a \times S^1] \cdot [a \times S^1] + 2[a \times S^1] \cdot [b \times \{1\}] + [b \times \{1\}] \cdot [b \times \{1\}] \\ &= 0 + 2[a] \cdot [b] + 0 = 2h \cdot [b] \end{aligned}$$

where the last intersection product is now taken in $H_*(M; \mathbb{Z})$. Now we have two possibilities:

1. h is a torsion element, so $2h \cdot [b]$ is a torsion element in $\mathbb{Z} = H_3(M; \mathbb{Z})$, therefore it is null; in this case $n' = n$;

2. h is not a torsion element; if we select the biggest integer d such that $h = dk$ for some $k \in H_1(M; \mathbb{Z})$, then we have $n' - n \equiv 0$ modulo $2d$.

This implies that in case the first obstruction α for two maps $f_0, f_1 : M \to S^2$ to be homotopic is null, we have constructed an object $\delta([f_0], [f_1]) = [n]$ which generalizes the classical notion of *Hopf number* and lives either in $\mathbb{Z}$ or in $\mathbb{Z}_{2d}$ according to the cases stated above. Of course if $[f_0] = [f_1]$ we can take Σ to be $L_0 \times [0, 1]$ so $\delta([f_0], [f_1]) = 0$; in other words δ is a second obstruction to homotopy. It is now not difficult to prove the following (we leave the details to the reader):

Theorem 6.2.7. *Two maps from M to S^2 are homotopic if and only if both the obstructions α and δ are null.*

Remark 6.2.8. In [52] on page 452 one can find a proof of the previous result in a more standard framework in obstruction theory.

6.3 Combinatorial realization of closed manifolds

In this section we exploit the results of Sections 5.2 and 6.2 to provide a calculus for 3-manifolds based on closed normal o-graphs. Since we are here disregarding further structures on manifolds we will refer to the unembedded notion of spine.

We address the reader for the notation and various comments on it to the beginning of Section 5.2, and also to Chapter 1. Let us just remind that $\mathcal{G} = \mathcal{G}_{\text{comb}}$ denotes the set of normal o-graphs of closed oriented manifolds with trivial bicoloration of the boundary. On $\mathcal{G}$ we have the standard sliding moves of Figures 1.4 and 1.5, and we already know that the coset space of $\mathcal{G}$ under the equivalence relation generated by these moves corresponds to the set of closed manifolds up to homeomorphism enhanced by vector fields up to homotopy and conjugation. The following result corresponds to Theorem 1.4.2.

Theorem 6.3.1. *The class $\mathcal{M}$ of closed, connected, oriented 3-manifolds up to oriented homeomorphism corresponds bijectively to the coset space of $\mathcal{G}$ under the equivalence relation generated by the moves of Figures 1.4 and 1.5 (standard sliding moves), and the move defined in Figure 1.6 (called combinatorial Pontrjagin move).*

Proof of 6.3.1. Recalling Theorem 5.2.1 (non-embedded version) and Theorem 6.2.2, we must show that the combinatorial Pontrjagin move for normal o-graphs translates the Pontrjagin move for vector fields of Definition 6.2.1. To be precise the theorem is deduced from the following facts which we shall establish:

I. On a manifold M, given vector fields up to homotopy $\overline{v}_0, \overline{v}_1$, where $\overline{v}_1$ is obtained from $\overline{v}_0$ via one Pontrjagin move, there exist elements G_0, G_1 of $\mathcal{G}$ which represent respectively $(M, \overline{v}_0)$ and $(M, \overline{v}_1)$ in the sense of Theorem 5.2.1, such that G_1 is obtained from G_0 via one combinatorial Pontrjagin move.

II. If G_0 and G_1 are elements of $\mathcal{G}$ related by one combinatorial Pontrjagin move then they define the same manifold.

Let us prove fact I. So we have, on a closed oriented manifold M, a vector field v_0 and a tube $T = D^2 \times S^1$ on which the induced field is $\partial/\partial\varphi$, with φ the coordinate of $S^1 = \mathbb{R}/2\pi\mathbb{Z}$. Then we construct a new field v_1 on M which coincides with v_0 outside T and is given on T by

$$v_1(\rho\, e^{i\vartheta}, e^{i\varphi}) = -\cos(\pi\rho)\frac{\partial}{\partial\varphi} - \sin(\pi\rho)\frac{\partial}{\partial\rho}.$$

What we will do is to find normal sections (in the sense of Section 5.2) for v_0 and v_1 which coincide outside T, and show that the evolution of the corresponding flow-spines is described in terms of normal o-graphs by the combinatorial Pontrjagin move.

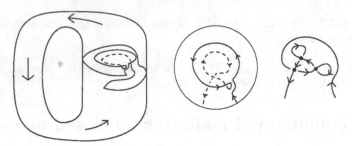

Figure 6.3: A section of the initial flow in the tube

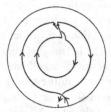

Figure 6.4: A section of the new flow in the tube

Let us consider inside T the discs Δ_0 and Δ_1 respectively described in Figures 6.3 and 6.4 (in the former we are showing a 3-dimensional picture, a cross-section view and a portion of normal o-graph which encodes the local picture; in the latter we are only showing the cross-section view). The following facts are obvious:

1. $\Delta_0 \cap \partial T = \Delta_1 \cap \partial T$;

2. For $i = 0, 1$ the disc Δ_i is transverse to v_i in T and meets all its half-orbits;

3. For $i = 0, 1$ there exists a normal section Σ_i of v_i on M such that $\Sigma_i \cap T = \Delta_i$, and moreover $\Sigma_0 \setminus T = \Sigma_1 \setminus T$;

4. Provided the portions of flow-spines in T associated to (Δ_0, v_0) and (Δ_1, v_1) have smooth components homeomorphic to discs, it is possible to find the above Σ_0 and Σ_1 so that the associated flow-spines are standard.

Let us now describe the portions of flow-spines carried by Δ_0 and Δ_1, showing that the standardness condition just stated is true and that the local picture can be faithfully encoded by a portion of normal o-graph. For Δ_0 this is obvious, and the o-graph was already shown in Figure 6.3.

For Δ_1 we will proceed with some care. To begin with, let us consider as a section of v_1 in T, instead of Δ_1, the union of a disc centred on the core of T and an annulus concentric to it. Then the situation is as simple as shown in Figure 6.5 (but of course this does not lead to a standard spine, and strictly speaking we do not have a section). Here (and in the sequel) a gray arc labeled α is the image of a black arc labeled α under the "first return function" associated to v_1 on the section under consideration. In Figures 6.6 and 6.7 we deal separately with the two non-trivial phenomena under which Δ_1 differs from the object of Figure 6.5. Note that here on the gray arcs we have trivalent vertices, which correspond to jumps of the "first return function" $\partial \Delta_1 \to \Delta_1$. With figures drawn like this, the normal o-graph is very easily obtained just by looking

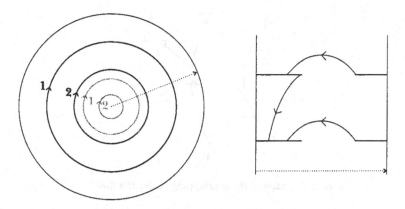

Figure 6.5: A simple "section" of the new flow

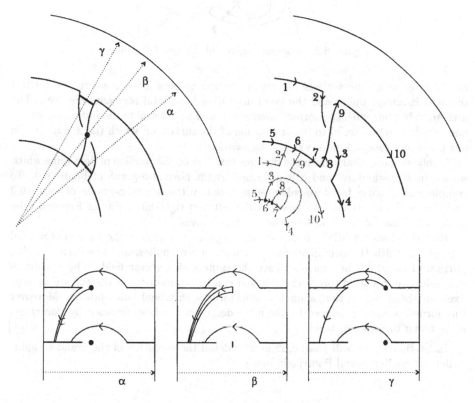

Figure 6.6: Image of the strip under the flow

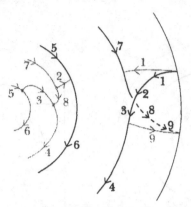

Figure 6.7: Image of the overlapping under the flow

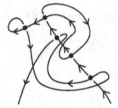

Figure 6.8: A normal o-graph of the new flow-spine

at the gray part: the vertices correspond to the objects like ⊥, where the vertical segment is always under and the two consecutive horizontal segments are over. The arcs already come with the correct orientation, and to determine the index of a vertex one only has to be careful to the orientation of the surface on which the ⊥ appears (in our particular case, all indices should be switched).

It follows quite clearly from the pictures that indeed the portion of flow-spine which we see in T is standard, and it is described by the normal o-graph of Figure 6.8. To complete the proof of fact I we only need to note that the normal o-graphs of Figure 6.3 and Figure 6.8 are related to those on the left and right-hand side of Figure 1.6 by means of planar isotopy and Reidemeister-type moves.

Fact II follows essentially from the same argument. Consider the portion of normal o-graph to which the combinatorial Pontrjagin move applies, and the corresponding portion of manifold, in which we have the spine and a vector field. The portion of manifold is just a solid torus, and the field can be homotoped on the torus (but kept fixed on the spine) so that a smaller solid torus with trivial flow appears. Moreover the normal o-graph produced by the move describes the flow obtained by Pontrjagin move along this smaller torus. $\boxed{6.3.1}$

In Section 7.5 we will analyze in greater detail the evolution of the branched spine under the combinatorial Pontrjagin move.

Chapter 7

Framed and spin manifolds

In this chapter we prove Theorems 1.4.3 and 1.4.4 concerning the combinatorial realizations of framed and spin 3-manifolds. For both the cases of framings and spin structures we start with some general topological results concerning how one can use a combing to construct the structure in exam, and then we show how the calculi already established can be refined to the new structures. A rather detailed account of the methods used has been given already in Chapter 1. We will avoid in this chapter to provide *embedded* versions of the results. In the first section we define a cochain representing the Euler class of the vector field carried by a branched spine; this cochain will play an essential role in the realization of framings and spin structures.

7.1 The Euler cochain

In this section we provide a recipe (inspired by [11]) to compute a cochain which represents the Euler class of the flow carried by a branched spine, directly from a normal o-graph. This cochain will play a central role in our treatment of framed and spin manifolds. In this section we put no restrictions on the boundary of the manifold and on the bicoloration induced on it by branched spines. So we will be using arbitrary normal o-graphs.

Proposition 7.1.1. *Consider a normal o-graph Γ of a branched standard spine P. The rules of Figure 7.1 associate to Γ a finite set of circuits such that:*

1. The circuits naturally correspond to the smooth components $\{\Delta_i\}$ of P;

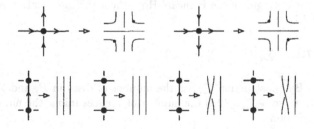

Figure 7.1: Computation of the Euler class

2. *On the circuit which corresponds to Δ_i there is an even number n_i of solid dots;*

3. *The Euler class of the vector field associated to the branched spine, as an element of $H^2(P; \mathbb{Z})$, is $[\sum_i (1 - n_i/2)\hat{\Delta}_i]$.*

Proof of 7.1.1. The first assertion is of general nature (Section 2.1, [6], [43]). The second and third assertion are proved simultaneously.

We note that, in a suitable sense, P has an oriented C^1-structure (Section 3.1), and then the cohomology class we are seeking is just the obstruction to the existence of a nowhere-zero tangent field to P. The rules of Figure 7.2 allow to construct a tangent

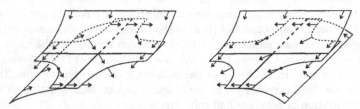

Figure 7.2: A tangent field to a branched spine

field near the vertices. Note that the circuits can be thought as curves in P cutting along which P splits into a regular neighborhood of $S(P)$ and discs corresponding to the regions $\{\Delta_i\}$. With this interpretation we see that the solid dots represent the points where the field is tangent to the circuits. It is easily seen that this field extends to the neighbourhood of $S(P)$ without adding any other point of tangency to the circuits.

If $n_i = 0$ we can extend the field on Δ_i with a zero of index 1. If $n_i > 0$ then $\partial\Delta_i$ is split into n_i segments on which the field points alternatively inside and outside Δ_i. Therefore we can extend the field with $n_i/2 - 1$ zeroes of index -1. $\boxed{7.1.1}$

We will denote by c_P the 2-cochain on P whose existence is guaranteed by the previous result.

In the previous proof we have used in the singular setting of branched spines arguments which a priori apply only to a smooth setting. We do not formally explain why this is possible, but at least we provide a supporting evidence by proving the following singular analogue of the Poincaré-Hopf theorem (the notation is the same as in Proposition 7.1.1):

Proposition 7.1.2. $\sum_i(1 - n_i/2) = \chi(P)$.

Proof of 7.1.2. By construction $\sum_i 1$ is the number of discs in P, and $\sum_i n_i/2$ is the number of vertices, so $-\sum_i n_i/2$ is the number of vertices minus the number of edges, whence the conclusion. $\boxed{7.1.2}$

We close this section by going back to the theme of Section 4.6 and showing that as an application of Proposition 7.1.1 we can obtain a more formal proof of Lemma 4.6.3. In fact, Figure 7.3 shows how to apply Proposition 7.1.1 to one of the MP-moves which change the flow. Locally there are before the move 9 portions of disc $\{\Delta_i\}_{i=1,\ldots,9}$ (in the figure i stands for Δ_i), and a new disc Δ_{10} appears after the move. As a direct

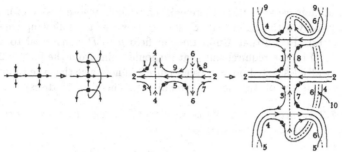

Figure 7.3: Evolution of the Euler class

application of Lemma 10.1.1 we see that locally H^2 is freely generated by $\hat{\Delta}_1, \hat{\Delta}_2, \hat{\Delta}_4, \hat{\Delta}_6$ and the other $\hat{\Delta}_i$'s are expressed in terms of them as follows:

$$\hat{\Delta}_3 = \hat{\Delta}_1 + \hat{\Delta}_2, \qquad \hat{\Delta}_5 = \hat{\Delta}_1 + \hat{\Delta}_2 - \hat{\Delta}_4, \quad \hat{\Delta}_7 = \hat{\Delta}_1 + \hat{\Delta}_2 + \hat{\Delta}_6 - \hat{\Delta}_4,$$
$$\hat{\Delta}_8 = \hat{\Delta}_1 + \hat{\Delta}_6 - \hat{\Delta}_4, \qquad \hat{\Delta}_9 = \hat{\Delta}_1 - \hat{\Delta}_4, \qquad \hat{\Delta}_{10} = \hat{\Delta}_4 - \hat{\Delta}_6.$$

Note that globally there could be relations between $\hat{\Delta}_1, \hat{\Delta}_2, \hat{\Delta}_4, \hat{\Delta}_6$. Now, with the move, the number of solid dots is unchanged on $\hat{\Delta}_1, \hat{\Delta}_2, \hat{\Delta}_3, \hat{\Delta}_6, \hat{\Delta}_8, \hat{\Delta}_9$, it is increased by 2 on $\hat{\Delta}_4, \hat{\Delta}_7$ and decreased by 2 on $\hat{\Delta}_5$, and there is no solid dot on $\hat{\Delta}_{10}$. Using Proposition 7.1.1 we see that the variation of $\mathcal{E}$ is given by:

$$-\hat{\Delta}_4 - \hat{\Delta}_7 + \hat{\Delta}_5 + \hat{\Delta}_{10} = -2\hat{\Delta}_6.$$

This computation is compatible with the fact that the comparison class of the new field with the old one is $-\hat{\Delta}_6$ (Lemma 4.6.3) and with Lemma 6.1.4. Actually we also deduce a new proof of Lemma 4.6.3: if we accept the fact that the comparison class carried by the move is expressed by a local formula, this formula will hold in particular when H^2 has a free $\mathbf{Z}$-summand generated by $\hat{\Delta}_6$, whence the conclusion.

7.2 Framings of closed manifolds

In this section we will describe the framework in which our combinatorial representation of framings is placed, and collect some general topological facts which will be used in the sequel. Throughout M will be a connected, closed, oriented 3-manifold and N will be the same manifold with an open 3-ball removed. Moreover P will be a branched standard spine of N with trivial bicoloration of the boundary 2-sphere and Γ will be the normal o-graph of P. We will think of P as being embedded in N in a C^1 way, with respect to the (singular) C^1-structure on P defined by the branching. We fix a vector field v on M carried by P, i.e. one which is positively transversal to P and traverses the ball complementary to N in the obvious way (recall that v is well-defined up to homotopy). We also define a Riemannian metric on M with respect to which v has unit length and is always orthogonal to P. For the sake of brevity we will call Euler class $\mathcal{E}$ of v the Euler class in $H^2(M; \mathbf{Z})$ of the distribution of 2-planes orthogonal to v. Recall that by the Mayer-Vietoris exact sequence M and N have the same homology and cohomology, with any coefficient group, up to order 2; moreover P is homotopy equivalent to N.

From general facts we know that there exists a framing whose first vector is homotopic to v if and only if the Euler class $\mathcal{E}$ is zero. However the following constructive proof will be useful. Recall that a unit tangent field μ to P transversal to $S(P)$ can be defined on $S(P)$ by the requirement that it should point from the locally 2-sheeted portion to the locally 1-sheeted portion (see Figure 7.2; note also that the Euler cochain c_P exactly represents the obstruction to extending this field to the discs).

Proposition 7.2.1. Let $x \in C^1(P; \mathbb{Z})$ be such that $\delta x = c_P$. Then there exists a unit vector field n_x on M with the following properties:

1. It is pointwise orthogonal to v (so it restricts to a tangent field to P);

2. It coincides with μ near the vertices of P;

3. On each edge e of P it is obtained by monotonically rotating μ of angle $-2\pi x(e)$ along the positive direction of e, as exemplified in Figure 7.4.

Moreover n_x is unique up to homotopy through fields satisfying the same properties. In particular the framing up to homotopy $f(x) = [(v, n_x)]$ is uniquely associated to x.

Figure 7.4: Modifying the tangent field to the spine along a singular edge

Proof of 7.2.1. We construct n_x inductively on the strata of M determined by P. So, let us denote by n_x a vector field on $S(P)$ tangent to P obtained as indicated in point 3 of the statement. We must first prove that n_x extends to a tangent field to P. To this end we need to recall some elementary facts.

Let $w : S^1 \to S^1$ be a map such that $w(z)$ is tangent to S^1 only at finitely many z's, all of which are transitions between an arc where w points inward D^2 and an arc where w points outwards. To each of these points we assign an index according to the rules of Figure 7.5. Then w extends to D^2 if and only if the sum of all the indices of tangency points is -1.

While proving that c_P represents $\mathcal{E}$ we have shown that each of the solid dots on the circuits corresponds to a tangency point of μ of index $-1/2$ on the corresponding

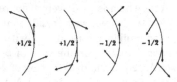

Figure 7.5: Indices of tangency points of a field on the unit circle

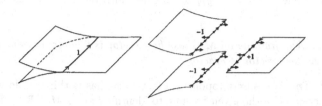

Figure 7.6: Indices of tangency points of the rotated field

disc. Now let us rotate μ by an angle -2π along an edge e; Figure 7.6 shows that on a disc Δ the sum of the indices of tangency points of the new field will be $-(\delta\hat{e})(\Delta)$. Therefore:

$$
\begin{aligned}
c(\Delta) &= 1 + \sum_{p\in\partial\Delta,\ p\ \text{vertex}} \text{ind}_{n_x}(p) \\
(\delta x)(\Delta) &= - \sum_{p\in\partial\Delta,\ p\ \text{not vertex}} \text{ind}_{n_x}(p) \\
\delta x = c &\ \rightarrow\ \sum_{p\in\partial\Delta} \text{ind}_{n_x}(p) = -1.
\end{aligned}
$$

This proves our assertion that n_x extends from $S(P)$ to a field normal to v on P. Then it also extends to N, which can be viewed as a small regular neighbourhood of P (we will call n_x again an extension to N). Now let us consider the closure D^3 of the ball complementary to N. If we identify TD^3 to $D^3 \times \mathbb{R}^3$ with v corresponding to a fixed vector u of $\mathbb{R}^3$, n_x restricted to ∂D^3 can be viewed as a map from the 2-sphere ∂D^3 to the unit circle S^1 in $u^\perp$. Since $\pi_2(S^1)$ is trivial then this map will indeed extend to the ball D^3.

Of course n_x is unique on $S(P)$ up to homotopy on $S(P)$. Its extension to P is unique up to homotopy on P because every map $\partial(D^2 \times [0,1]) \to S^1$ extends to a map $D^2 \times [0,1] \to S^1$, and then the extension to M is also unique up to homotopy because every map $\partial(D^3 \times [0,1]) \to S^1$ extends to a map $D^3 \times [0,1] \to S^1$. $\boxed{7.2.1}$

We will assume from now on that the Euler class $\mathcal{E}$ is indeed zero (our results would be true but empty otherwise). In the sequel we will denote by $(x)_2$ the reduction modulo 2 of a $\mathbb{Z}$-cochain x.

Proposition 7.2.2. *If $x, y \in C^1(P; \mathbb{Z})$ and $\delta x = \delta y = c_P$ then $f(x) = f(y)$ if and only if $(x - y)_2$ is a $\mathbb{Z}_2$-coboundary.*

Proof of 7.2.2. If we use the first framing to identify the tangent bundle TM to $M \times \mathbb{R}^3$, the second framing will be represented by a map $\alpha : M \to SO(3)$. Since the two framings have the same first vector, we will have $\alpha(x)e_1 = e_1$ for all $x \in M$, where e_1 is the first vector of the canonical basis of $\mathbb{R}^3$. Now it is very easy to check that a map α with this property is homotopic to the identity if and only if $\alpha_* : \pi_1(M) \to \pi_1(SO(3))$ is the zero homomorphism (if α_* is zero then α lifts to the universal cover S^3 of $SO(3)$ to a map which is certainly not surjective).

Now $\pi_1(SO(3)) = \mathbb{Z}_2$, so α_* can be canonically viewed as an element of $H^1(M; \mathbb{Z}_2)$. Recalling that $\pi_1(SO(3))$ is generated by the rotation in the plane orthogonal to e_1, it is not difficult to check that $(x - y)_2$ is a cocyle which represents α_* as an element of $H^1(M; \mathbb{Z}_2)$.

$$\boxed{7.2.2}$$

Lemma 7.2.3. *Any framing on M whose first vector is homotopic to v can be homotoped so that the first vector is exactly v.*

Proof of 7.2.3. This is a purely topological fact which has nothing to do with the rest of our construction. Use the given framing to identify TM to $M \times \mathbb{R}^3$. Then what we have is a homotopy $M \times [0,1] \ni (x,t) \mapsto v_t(x) \in S^2$ with $v_0(x) = e_1$ for all $x \in M$, which we would like to lift to $M \times [0,1] \to SO(3)$ with the identity at time 0, with respect to the mapping $SO(3) \ni A \mapsto Ae_1 \in S^2$. The conclusion that this can be done follows from the fact that the map $SO(3) \to S^2$ defines a locally trivial fibration, so it has the homotopy lifting property with respect to compact manifolds.

$$\boxed{7.2.3}$$

Proposition 7.2.4. *Any framing on M whose first vector is homotopic to v belongs to a homotopy class of framings $f(x)$ for some $x \in C^1(P; \mathbb{Z})$ with $\delta x = c_P$.*

Proof of 7.2.4. According to the previous lemma it is sufficient to refer to framings whose first vector is v itself. Let us fix one such framing carried by some $x_0 \in C^1(M; \mathbb{Z})$ with $\delta x_0 = c_P$, and use it to identify TM with $M \times \mathbb{R}^3$. Then a vector field n orthogonal to v will be given by a map from M to the unit circle S^1 in the plane $e_1^\perp$. Moreover the homotopy class of the framing (v, n) is determined by the homotopy class of n as a map $M \to S^1$ (not quite the contrary, according to Proposition 7.2.2). Now, the set $\pi(M, S^1)$ of homotopy classes of maps $M \to S^1$ can be identified to $H^1(M; \mathbb{Z})$. Let us consider the function which assigns to a cocycle $x \in Z^2(P; \mathbb{Z})$ the class $[n_{x+x_0}] \in \pi(M, S^1)$. Then it is quite easy to see that this function induces the identity map of $H^1(P; \mathbb{Z}) \cong H^1(M; \mathbb{Z})$ under the identification $H^1(M; \mathbb{Z}) \cong \pi(M, S^1)$, and the conclusion follows.

$$\boxed{7.2.4}$$

We still need the following elementary fact.

Lemma 7.2.5. *If $x, y \in C^1(P; \mathbb{Z}_2)$, $x - y \in B^1(P; \mathbb{Z}_2)$ and there exists $\tilde{x} \in C^1(P; \mathbb{Z})$ with $x = (\tilde{x})_2$ and $\delta \tilde{x} = c_P$, then there also exists $\tilde{y} \in C^1(P; \mathbb{Z})$ with $y = (\tilde{y})_2$ and $\delta \tilde{y} = c_P$.*

Proof of 7.2.5. If $x - y = \delta z$ with $z \in C^0(P; \mathbb{Z}_2)$, choose any $\tilde{z} \in C^0(P; \mathbb{Z})$ with $z = (\tilde{z})_2$, and set $\tilde{y} = \tilde{x} - \delta \tilde{z}$.

$$\boxed{7.2.5}$$

What we have stated so far in this section is sufficient to develop the framing calculus as we will do below. However for the sake of completeness we also provide a very general description of homotopy classes of framings on M, by means of the algebraic topology of M. We fix a certain framing of M, so that the set of homotopy classes of framings can be identified with the set $\pi(M, SO(3))$. Note that this set can be naturally endowed with a group structure induced by pointwise multiplication. We denote by $p : S^3 \to SO(3)$ the universal cover.

Theorem 7.2.6. *There exists an exact sequence of group homomorphisms*

$$0 \to \mathbb{Z} \overset{\phi}{\longrightarrow} \pi(M, SO(3)) \overset{\psi}{\longrightarrow} H^1(M; \mathbb{Z}_2) \to 0,$$

where $\phi(n) = [p \circ f_n]$, $f_n : M \to S^3$ being any map of degree n, and $\psi([f]) = f_*$ under the identification $H^1(M; \mathbb{Z}_2) \cong \mathrm{Hom}(\pi_1(M), \mathbb{Z}_2)$.

Proof of 7.2.6. Since $\pi(M, S^3) \cong \mathbb{Z}$ via degree, ϕ is well-defined, and of course ψ also is. Let us prove exactness.

If $p \circ f_n$ is homotopic to a constant map then we can lift the homotopy to S^3, proving that f_n is null-homotopic, so $n = 0$. Therefore ϕ is injective.

Given $f : M \to SO(3)$ we have that $\psi([f]) = 0$ if and only if f lifts to S^3, and we immediately deduce that the kernel of ψ is the image of ϕ.

We are left to prove that ψ is surjective. To this end consider a standard spine P of N (the argument could be adapted to any cellularization of M, but we stick to our favourite objects). We first prove that every homomorphism $\alpha : \pi_1(P) \to \mathbb{Z}_2$ is induced by some map $j : P \to \mathbb{P}^2(\mathbb{R})$. We denote by $a : [0, 1] \to \mathbb{P}^2(\mathbb{R})$ a simple loop which generates $\pi_1(\mathbb{P}^2(\mathbb{R}))$. We choose a maximal tree T in the singular set of P, and define j to be constantly equal to the basepoint of a on T. On the other edges of T we define j to be constant or equal to a, according to the value of α on the element of $\pi_1(P)$ which corresponds to the edge. On the boundary of each disc of P there will be an even number of edges on which j is non-constant, and hence j extends to the disc. Now to conclude we think of $\mathbb{P}^2(\mathbb{R})$ as a subset of $\mathbb{P}^3(\mathbb{R}) \cong SO(3)$, and all we have to do is to extend j from P to M. Of course j extends to N, so we are left to extend a map $S^2 \to \mathbb{P}^3(\mathbb{R})$ to D^3. To this end we only need to note that the second homotopy group of $\mathbb{P}^3(\mathbb{R})$ is trivial, because this is the case for the universal cover S^3. $\boxed{7.2.6}$

Remark 7.2.7. Assume we have a branched standard spine P of N which carries a vector field v on M, and let us fix a framing f_0 whose first vector is v. We use f_0 to identify the set of framings on M with $\pi(M, SO(3))$, and we consider the subset F_v of framings whose first vector is homotopic to v. We know that very element f of F_v corresponds to a certain $x_f \in C^1(P; \mathbb{Z})$ with $\delta x_f = c_P$. Then the map $F_v \mapsto [(x_f - x_{f_0})_2]$ is the restriction of ψ to F_v, and it is in general not surjective. Its image consists precisely of the homomorphisms $\pi_1(M) \to \mathbb{Z}_2$ which are induced via reduction modulo 2 by homomorphisms $\pi_1(M) \to \mathbb{Z}$.

7.3 The framing calculus

This section is devoted to the proof of Theorem 1.4.3. We first outline the strategy and then turn to a punctual description. As in the previous section N will be a connected, oriented, compact manifold bounded by S^2 and M will be the associated compact manifold. If P is a branched standard spine of N with trivial bicoloration on ∂N we denote by v_P a vector field on M positively transversal to P and by $c_P \in C^2(P; \mathbb{Z})$ the preferred representative of the Euler class of v_P.

Consider the set of all pairs (P, x), where P is as above and $x \in C^1(P; \mathbb{Z}_2)$ admits a lifting $\tilde{x} \in C^1(P; \mathbb{Z})$ such that $\delta \tilde{x} = c_P$. According to Proposition 7.2.1 to every such pair (P, x) we can associate a framing $f(P, x)$. Moreover, since we know that every vector field up to homotopy is carried by some P, using Proposition 7.2.4 we see that all framings of M can be obtained as $f(P, x)$ for some (P, x). By Proposition 7.2.2, considering a fixed P, we have that $f(P, x) = f(P, y)$ if and only if $x - y \in B^1(P; \mathbb{Z}_2)$; this relation can be encoded by the "local" move on pairs $(P, x) \to (P, x + \delta \hat{v})$, where v is a vertex of P (it is important to note that by Lemma 7.2.5 we do not need to verify the lifting property for the new cochain $x + \delta \hat{v}$).

Now we want to allow also P to change. The idea is just that for any of the sliding moves $P \to P'$ of the calculus for combed manifolds, we should find a general recipe to lift the move to a move of pairs $(P, x) \to (P', x')$, where $f(P, x) = f(P', x')$.

Remark 7.3.1.   1. For the self-sliding move (locally from 0 to 2 vertices), we can imagine that the cochain x (or, better to say, its lift $\tilde{x}$) affects the vector field tangent to P only far from where the sliding takes place;

2. For all other sliding moves (locally from 2 to 3 vertices) up to adding some $\delta\hat{v}$ we can assume that the dual of the edge which is blown up by the move has coefficient 0 in x, and we can assume that x affects the tangent field to P only far from the sliding area.

According to this fact for each of the sliding moves we will have only one case to consider.

Having outlined the proof, we now turn to a more detailed explanation of the combinatorial statement given. We first note that the objects $\mathcal{G}_{\text{fram}}$ of our representation naturally correspond to pairs (P, x) as stated, because:

- Axiom F1 defines a class $x \in C^1(P; \mathbb{Z}_2)$ as the linear combination of the duals of the edges in which the coefficient of the dual of an edge is the colour of that edge.

- Axiom F2 translates the fact that there should be a lifting $\tilde{x} \in C^1(P; \mathbb{Z})$ of x with $\delta\tilde{x} = c_P$.

Concerning axiom F2 we also want to note that the orientation of the circuits is different in Figure 1.3 from the orientation used so far (in particular in Chapter 3). The reason is that here we are orienting the circuits as boundaries of the discs, whereas in Chapter 3 we were orienting them as boundaries of the neighbourhood of the singular set.

The first of the moves shown in Figure 1.7 of course translates the move $(P, x) \mapsto (P, x + \delta\hat{v})$. We must show now that the other moves correspond to enhancements to pairs (P, x) of the sliding moves for normal o-graphs. To see this we will use Remark 7.3.1. Moreover we will use planar figures only, with the convention that near the sliding region the spine should be almost horizontal and the tangent field to it should be obtained as the pull-back of a horizontal field under the projection on the horizontal plane. So in the various pictures we will have a certain directed planar graph with crossings, and a planar field represented by some of its integral curves. At crossings, this field should locally point to the right of both branches of the graph, and it should have isolated tangency points to edges. Moreover a coefficient 1 on an edge will result from a pair of tangency points to the edge. Instead of completely formalizing these conventions, we put them at work on the easiest case of the self-sliding move, which is faced in Figure 7.7, and corresponds to the second move in Figure 1.7.

Now we have to deal with the other slidings. The first case we examine is when both slidings are possible. This is done in Figure 7.8, which translates into the moves of the first line of Figure 1.8.

Note that instead of the colouring which directly arises from Figure 7.8 we have used in some cases an equivalent one. The various different cases are obtained by obvious symmetries from the case explicitly described.

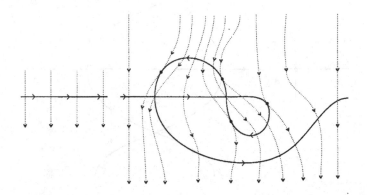

Figure 7.7: Homotopies of horizontal fields

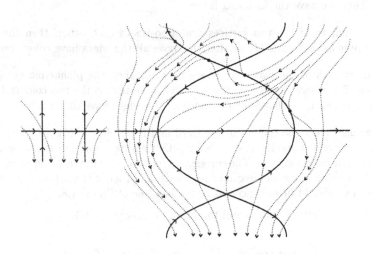

Figure 7.8: More homotopies of horizontal fields

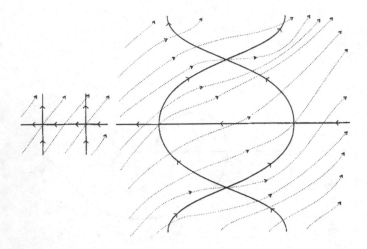

Figure 7.9: Still more homotopies of horizontal fields

The second case we deal with is when the ends of the edge which is blown up are either both above or both below, but only one sliding is possible. This is done in Figure 7.9, which translates into the moves of the second line in Figure 1.8 (again, use symmetries to get all possible cases).

We next consider the case when one of the ends of the blown up edge is above, and the other end is below. Up to symmetry we can restrict to the case where the first end is below. Then we have the following facts:

(a) If the indices of the two vertices are opposite to each other, then the planar picture is as in Figure 7.9, so after the move all the edges have colour zero.

(b) If the indices of the two vertices are both -1 then the planar picture is as in Figure 7.8, where the upper crossing is not a vertex, so the two colours 1 cancel and again we have that after the move all the edges have colour zero.

These conclusions are the content of the third and fourth line of Figure 1.8.

We are left to examine the case where the first end is below and the second one is above, and both indices are $+1$. The corresponding planar picture is again essentially as shown in Figure 7.8 (with all directions reversed), and then it is quite easy to conclude that the sliding should be enhanced as in the last line of Figure 1.8.

This concludes our argument and the proof of Theorem 1.4.3.

7.4   Spin structures on closed manifolds

Let all the notations P, N, M, v_P, c_P listed at the beginning of Section 7.3 be fixed, without the assumption that $[c_P]$ should be null in $H^2(P; \mathbb{Z})$. Let us recall that spin structures on M correspond bijectively to framings of M on $S(P)$ which have the property of extending to P and are regarded up to homotopy on $S(P)$.

Lemma 7.4.1. *Every spin structure on M can be represented by a framing of M on $S(P)$ whose first vector is v_P.*

Proof of 7.4.1. We fix some framing of M on $S(P)$ with first vector v_P, so that any spin structure is represented by some $s : S(P) \to SO(3)$, and we want to homotope every such s on $S(P)$ to another $s' : S(P) \to SO(3)$ with $s'(\cdot)e_1 \equiv e_1$. Up to a small perturbation we can assume that $s(\cdot)e_1$ is never antipodal to e_1. Let us denote by $\gamma(x) : [0,1] \to SO(3)$ the shortest path of rotations in the $(e_1, s(x)e_1)$-plane with $\gamma_0(x) = \mathrm{Id}$ and $\gamma_1(x)s(x)e_1 = e_1$, parametrized with constant speed. Then the map $(x, t) \mapsto \gamma_t(x)s(x)$ is a homotopy between s and a framing s' as required. $\boxed{7.4.1}$

The following result gives a constructive proof of the well-known fact that spin structures on M are an affine space over $H^1(P; \mathbb{Z}_2)$. The proof is just outlined.

Proposition 7.4.2. *Spin structures on M correspond bijectively and constructively to cochains $z \in C^1(P; \mathbb{Z}_2)$ such that $\delta z = (c_P)_2$, where two cochains are viewed as equivalent if they differ by a 1-coboundary.*

Proof of 7.4.2. Let μ be the tangent field to P along $S(P)$ which points from the locally 2-sheeted side to the locally 1-sheeted side. We perturb μ exactly as we did in Proposition 7.2.1, using any lifting to $\mathbb{Z}$ of the given $\mathbb{Z}_2$-cochain. Call w the resulting field. We must show that the framing (v_P, w) extends to every disc Δ. Assume by simplicity that Δ is embedded, and trivialize the tangent bundle to M on Δ using v_P as first vector. Using the assumption that $\delta z = (c_P)_2$ it is quite easy to see that the framing (v_P, w) is represented by a map

$$S^1 \cong \partial\Delta \ni x \mapsto \begin{pmatrix} 1 & 0 \\ 0 & \zeta(x) \end{pmatrix} \in SO(3),$$

where $\zeta : S^1 \to S^1$ has even degree. Therefore the framing extends to Δ.

Using the previous lemma and the easy fact that any $\mathbb{Z}$-lifting of z will lead to the same framing up to homotopy on $S(P)$, we see that all spin structures arise from the previous construction. Moreover, using the fact that $\pi_2(SO(3)) = 0$, we see that two framings of M on P are homotopic on P if and only if their restrictions to $S(P)$ are homotopic on $S(P)$. From this, with the same argument used in the proof of Proposition 7.2.2, one sees that two cochains define the same spin structure if and only if they differ by a coboundary. $\boxed{7.4.2}$

7.5   The spin calculus

In this section we will establish Theorem 1.4.4. According to Proposition 7.4.2, every element of $\mathcal{M}_{\mathrm{spin}}$, i.e. a closed 3-manifold with a certain spin structure, is represented by a pair (P, z), where P is a branched standard spine of the punctured manifold with trivial bicoloration on the boundary, and $z \in C^1(P; \mathbb{Z}_2)$ is such that $\delta z = (c_P)_2$. Going back to the definition of c_P in Section 7.1 and recalling the correspondence between branched standard spines and normal o-graphs, one sees very easily that there is a natural bijection between the set of pairs (P, z) as stated and the set $\mathcal{G}_{\mathrm{spin}}$ of spin normal o-graphs defined in Chapter 1.

Proposition 7.4.2 also implies that, given two elements of $\mathcal{G}_{\text{spin}}$ which differ only for the $\mathbb{Z}_2$-colorings, they define the same element of $\mathcal{M}_{\text{spin}}$ if and only if they are related by the move on the left-hand side on Figure 1.7. Again the same proposition implies that given a spin manifold and a spine P as above of the manifold, there exists some z as above such that (P, z) represents the spin structure. Therefore to complete the spin calculus we only need to take into account the moves of the closed calculus (Theorem 1.4.2) and refine them to moves which preserve the spin structures carried by $\mathbb{Z}_2$-colorings of the edges.

The case of standard sliding moves is easy. Since a spin structure is a trivialization of the tangent bundle on the 2-skeleton, the local pictures to take into consideration are exactly those already envisaged in Section 7.3. Therefore the spin-refinement of standard sliding moves is exactly the same as for the framing calculus.

We are left to deal with the combinatorial Pontrjagin move. In Figure 7.10 we analyze in detail the situation before the move: locally there are 3 vertices u_1, u_2, u_3, 7 edges (5 of which, denoted a_i for $i = 1, \ldots, 5$, are completely contained in the local picture) and 5 discs, denoted Δ_i for $i = 1, \ldots, 5$ (note that we write i for Δ_i). Only the discs Δ_4 and Δ_5 are completely contained in the local picture. The contribution of the local picture to $(c)_2$ is $\hat{\Delta}_1 + \hat{\Delta}_3 + \hat{\Delta}_4 + \hat{\Delta}_5$ (this means, for instance, that the coefficient of $\hat{\Delta}_3$ in $(c)_2$ is 1 if there are no other marked points on $\partial \Delta_3$ outside the local picture. Since $\partial \Delta_4 = a_2$ and $\partial \Delta_5 = a_4$, if a cochain z is such that $\delta z = (c)_2$ then necessarily $\hat{a}_2$ and $\hat{a}_4$ have coefficient 1 in z. Moreover up to adding δu_2 and/or δu_3 we can assume that $\hat{a}_1$ and $\hat{a}_5$ have coefficient 0 in z. Therefore the local possibilities for z are the following:

$$\begin{align}
(1) \qquad & \hat{a}_2 + \hat{a}_4 \\
(2) \qquad & \hat{a}_2 + \hat{a}_3 + \hat{a}_4;
\end{align}$$

Correspondingly the local contributions to $(c)_2 + \delta z$ are:

$$\begin{align}
(1) \qquad & \hat{\Delta}_1 + \hat{\Delta}_3 \\
(2) \qquad & \hat{\Delta}_1 + \hat{\Delta}_2 + \hat{\Delta}_3.
\end{align}$$

Let us turn to the situation after the move, described in Figure 7.11. Note that there are 5 vertices v_i, 11 edges (9 of which, denoted b_i, completely within the local picture) and 7 discs Σ_i (4 of which inside the local picture; again we are writing i for Σ_i). The local contribution to $(c)_2$ is given by

$$\hat{\Sigma}_1 + \hat{\Sigma}_2 + \hat{\Sigma}_3 + \hat{\Sigma}_7.$$

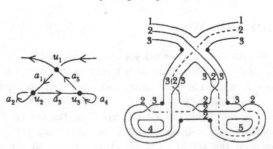

Figure 7.10: Spin calculus: situation before the Pontrjagin move

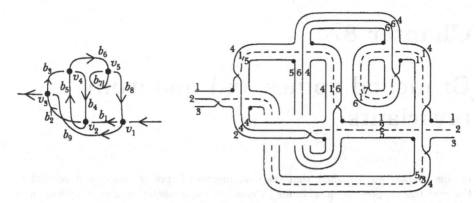

Figure 7.11: Spin calculus: situation after the Pontrjagin move

We want to determine which cochains z' are such $\delta z' = (c)_2$ locally. First of all since $\partial\Sigma_7 = b_7$, the coefficient of $\hat{b}_7$ in z' must be 1. Moreover up to adding δv_4, δv_2 and δv_5 we can assume that $\hat{b}_3, \hat{b}_4$ and $\hat{b}_6$ have coefficient 0 in z' (note that we are not using δv_1 and δv_3, to avoid modifying the coefficients outside the local picture). Now we have:

$$\delta\hat{b}_1 = \hat{\Sigma}_1 + \hat{\Sigma}_2 + \hat{\Sigma}_5 \qquad \delta\hat{b}_2 = \hat{\Sigma}_2 \qquad \delta\hat{b}_5 = \hat{\Sigma}_4 + \hat{\Sigma}_5 + \hat{\Sigma}_6$$
$$\delta\hat{b}_7 = \hat{\Sigma}_1 + \hat{\Sigma}_6 + \hat{\Sigma}_7 \qquad \delta\hat{b}_8 = \hat{\Sigma}_4 \qquad \delta\hat{b}_9 = \hat{\Sigma}_3 + \hat{\Sigma}_4 + \hat{\Sigma}_5.$$

We are led to the following system of equations in $\alpha_1, \alpha_2, \alpha_5, \alpha_8, \alpha_9$:

$$\delta(\alpha_1\hat{b}_1 + \alpha_2\hat{b}_2 + \alpha_5\hat{b}_5 + \hat{b}_7 + \alpha_8\hat{b}_8 + \alpha_9\hat{b}_9)(\Sigma_j) = 0, \qquad j = 4, 5, 6.$$

It is not difficult to check that the solutions give exactly the following locally admissible possibilities for z':

(1')	$\hat{b}_1 + \hat{b}_5 + \hat{b}_7 + \hat{b}_8$
(2')	$\hat{b}_1 + \hat{b}_2 + \hat{b}_5 + \hat{b}_7 + \hat{b}_8$
(3')	$\hat{b}_5 + \hat{b}_7 + \hat{b}_9$
(4')	$\hat{b}_2 + \hat{b}_5 + \hat{b}_7 + \hat{b}_9$;

Correspondingly the local contributions to $(c)_2 + \delta z$ are:

(1')	$\hat{\Sigma}_1 + \hat{\Sigma}_3$
(2')	$\hat{\Sigma}_1 + \hat{\Sigma}_2 + \hat{\Sigma}_3$
(3')	$\hat{\Sigma}_2$
(4')	0.

Since for $j = 1, 2, 3$ the discs Δ_j and Σ_j correspond to each other, we see that the local possibilities (3') and (4') for z' are not compatible with any local possibility for z, whereas (1') corresponds to (1) and (2') corresponds to (2).

It is now easy to see that the first row in Figure 1.9 corresponds to the Pontrjagin move with the spin colourings (1)-(1'), and the second row to colourings (2)-(2'). This concludes the proof of Theorem 1.4.4.

Chapter 8

Branched spines and quantum invariants

In this chapter we will show that the environment of spin calculus, as developed in the previous chapter, is suitable for an effective implementation and computation of the Turaev-Viro invariants in their spin-refined version. Moreover we will illustrate, using our machinery, the various refined versions of the Turaev-Walker theorem. In [7] we have shown that o-graphs can be used to encode the basic version of Turaev-Viro invariants and support the proof of the basic version of the Turaev-Walker theorem, so most of the present chapter can be viewed as a generalization of [7].

Our methods are based on the approach of J. Roberts [47], [48] to the Turaev-Viro theory; this approach exploits Lickorish's original use of skein recoupling theory in connection with invariants (see [34], [28], [3], [4]). It is by now well-known that skein recoupling theory allows to construct, in a self-contained and elementary way, the quantum invariants of 3-manifolds corresponding to the special case of quantum deformations of $sl(2;\mathbb{C})$. It is conceivable that, with suitable technical variations, what follows could be carried over to the level of generality axiomatized by Turaev in [56].

We emphasize that this chapter contains essentially no new ideas with respect to [48], [3] (and others), except for some cues in Section 8.3 and particularly in Section 8.4. The main reason for including it is that it provides a first application of our techniques.

8.1 More on spin structures

In this chapter all homology and cohomology groups have coefficients in $\mathbb{Z}_2$, and we will always omit the group from the notation.

Let M be a closed oriented 3-manifold. We will denote by $\mathrm{Spin}(M)$ the set of spin structures on M (an affine space over $H^1(M)$). We will denote by $-M$ the same manifold with opposite orientation. Then we have a natural bijection $s \mapsto -s$ of $\mathrm{Spin}(M)$ onto $\mathrm{Spin}(-M)$ obtained by considering s as a positive framing of M over the 1-skeleton, and multiplying it (thought as a triple of vector fields) by -1 so to get a positive framing $-s$ on the 1-skeleton of $-M$.

Now, since the 3-ball and the punctured 3-ball both have a unique spin structure, one easily sees that

$$\mathrm{Spin}(M\#(-M)) \cong \mathrm{Spin}(M) \times \mathrm{Spin}(-M),$$

and moreover the following map is a bijection:

$$\Phi : \text{Spin}(M) \times H_2(M) \;\to\; \text{Spin}(M\#(-M)) \cong \text{Spin}(M) \times \text{Spin}(-M)$$
$$(s,c) \;\mapsto\; (s, -s + D(c))$$

(here, and always in the sequel, D denotes Poincaré duality). In particular we can restrict Φ to $\text{Spin}(M) \times \{0\}$ to get a diagonal inclusion $\varphi : \text{Spin}(M) \hookrightarrow \text{Spin}(M\#(-M))$.

Assume now that L is an n-component framed link in S^3 which represents by surgery some closed oriented 3-manifold Q. Recall that in such a case Q is the boundary of the 4-manifold W_L obtained by attaching 2-handles to D^4 along the components L_i of L. These 2-handles give a basis of $H_2(W_L)$ and, with a slight abuse of notation, will still be denoted by L_i. We will denote by $\mathcal{K}(L)$ the set of characteristic elements of $H_2(W_L)$, i.e. those elements k such that for all x in $H_2(W_L)$ one has $x \cdot x = k \cdot x$ (here we are using the homology intersection product). The set $\mathcal{K}(L)$ can be characterized as follows: let us denote by B_L the linking matrix of L (with the framing p_i of L_i in position (i,i), as usual); then k is characteristic if and only if it has the form $k = \sum_{i=1}^n c_i L_i$, where $c \in (\mathbb{Z}_2)^n$ is a solution of the $\mathbb{Z}_2$-linear system $B_L c \equiv p$ modulo 2. It is known that there exists a bijection

$$\Psi_L : \text{Spin}(Q) \to \mathcal{K}(L)$$

where $\Psi_L(s)$ is defined as the dual to the element $w_2(s) \in H^2(W_L, Q)$ which represents the obstruction to extending s from Q to W_L. We conclude by remarking that in case $Q = M\#(-M)$ for some closed oriented M, by composition of Ψ_L with the map φ defined above we realize $\text{Spin}(M)$ as a subset of $\mathcal{K}(L)$.

8.2 A review of recoupling theory and Reshetikhin-Turaev-Witten invariants

Concerning recoupling theory we will confine ourselves here to the facts we strictly need for our purposes, addressing the reader to [34], [28] and [3] for further details.

Let us fix an integer $r \geq 3$. Starting from the Kauffman bracket with the value $A = \exp(2\pi i/4r)$ of the fundamental parameter, recoupling theory assigns a value (a complex number) to each admissibly coloured trivalent ribbon in S^3 (in particular, to each framed link). Here an admissible coloration is obtained by attaching to each edge of the ribbon (or component of the link) an integer in $\{0, 1, 2, \cdots, r-2\}$, in such a way that if a, b, c are the colours of the edges incident to a trivalent vertex then $a + b + c$ is even and there exists a triangle in the Euclidean plane with edges of lengths a, b, c. Note that a framed link has no vertices, so every coloration of the components with elements of $\{0, 1, 2, \cdots, r-2\}$ will be admissible.

If the components of a framed link are coloured by formal linear combinations of admissible colours one can still define the evaluation, extending the definition by linearity. Let us denote by Δ_a (the a-th Chebyschev polynomial computed in A) the evaluation of the 0-framed unknot coloured by a. Let us also choose a complex number $\eta \neq 0$ such that $\eta^{-2} = \sum_{a=0}^{r-2} \Delta_a^2$, and define a formal colour Ω as the linear combination of colours in which each a has coefficient $\eta \Delta_a$. We also define Ω_0 (resp. Ω_1) by considering in the same linear combination only even (resp. odd) colours; hence $\Omega = \Omega_0 + \Omega_1$. Finally we define κ (resp. κ_0) as the evaluation of the 0-framed unknot

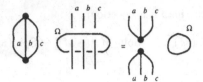

Figure 8.1: The fusion rule

coloured by Ω (resp. Ω_0). In general if L is a framed link we will denote by $\langle\langle L \rangle\rangle$ the evaluation of L with all components coloured by Ω; if z is a rule which assigns to every component of L a symbol 0 or 1, we will denote by $\langle\langle L \rangle\rangle_z$ the evaluation of L where each component K of L is coloured by $\Omega_{z(K)}$. All these definition will be used in the sequel to construct the invariants.

We conclude by recalling one of the (very many) remarkable properties of recoupling theory, namely the fusion rule illustrated in Fig. 8.1 (this will also be mentioned later in this chapter).

RTW invariants. Consider a closed oriented 3-manifold Q represented by surgery along a framed link L in S^3. Given $r \geq 3$, the absolute version of the invariant predicted by Witten [61] and constructed by Reshetikhin and Turaev [46] can be expressed, according to Lickorish [34] as

$$\mathrm{RTW}^r(Q) = \eta \kappa^{-\sigma(L)} \langle\langle L \rangle\rangle$$

where $\sigma(L)$ is the signature of the linking matrix B_L of L. The invariant is normalized in such a way that the following relations hold:

- $\mathrm{RTW}^r(-Q) = \overline{\mathrm{RTW}^r(Q)}$;

- $\mathrm{RTW}^r(Q_1 \# Q_2) = \eta^{-1}\mathrm{RTW}^r(Q_1)\mathrm{RTW}^r(Q_2)$.

Blanchet [3] (see also [31], where the techniques are different) has defined invariants $\mathrm{RTW}^r(Q, s)$ when $r \equiv 0$ modulo 4 and s is a spin structure on Q, and $\mathrm{RTW}^r(Q, x)$ when $r \equiv 2$ modulo 4 and $x \in H^1(Q)$. (Actually Blanchet's definition works for more general roots of unity, but he proves that these are the truly relevant cases.) Let us outline how this definition goes, starting from the case $r \equiv 2$ modulo 4.

Recall that the elements of $H^1(Q)$ correspond to solutions $c \in (\mathbb{Z}_2)^n$ of the homogeneous system $B_L c \equiv 0$ modulo 2. Let $c(x)$ correspond to the given x; we can view $c(x)$ as a function which assigns the symbol $c_i(x) \in \mathbb{Z}_2$ to the component L_i of L. Then, using the definitions given above, the invariant is defined as

$$\mathrm{RTW}^{r(2)}(Q, x) = \eta \kappa_0^{-\sigma(L)} \langle\langle L \rangle\rangle_{c(x)}.$$

In case $r \equiv 0$ modulo 4 we can proceed in a similar way. Given a spin structure s on Q one considers (see Section 8.1) the corresponding element $\Psi_L(s) \in \mathcal{K}(L) \subset H_2(W_L)$, which again can be viewed as a colouring rule for the components of L, and then set

$$\mathrm{RTW}^{r(0)}(Q, s) = \eta \kappa_0^{-\sigma(L)} \langle\langle L \rangle\rangle_{\Psi_L(s)}.$$

In both cases the proof of invariance relies on a refinement of the Kirby calculus to pairs (Q, x) or (Q, s).

8.3 Turaev-Viro invariants

We start by defining the absolute version of the invariant, following the approach of Roberts (later we will mention the original definition). We have seen in Section 2.3 (Definition 2.3.2) that to an o-graph Γ representing a 3-manifold N with boundary one can associate a framed link $L(\Gamma)$ in S^3 which represents by surgery the mirrored manifold $N \sqcup_\partial (-N)$. (The definition of $L(\Gamma)$ is actually ambiguous, but this will do no harm.) In the special case where $\partial N = S^2$ and $M = \widehat{N}$ we have $N \sqcup_\partial (-N) = M\#(-M)$. Therefore it is obvious that if we define

$$\mathrm{TV}^r(M) = \eta^2 \kappa^{-\sigma(L(\Gamma))} \langle\!\langle L(\Gamma) \rangle\!\rangle$$

we get an invariant of M, which coincides with $\eta \mathrm{RTW}^r(M\#(-M))$. The properties of RTW^r then imply that $\mathrm{TV}^r(M) = |\mathrm{RTW}^r(M)|^2$. More in general one can find in [7] the definition of $\mathrm{TV}^r(N)$ when N has a boundary, and the proof that $\mathrm{TV}^r(N) = \eta^{-\chi(N)} \mathrm{RTW}^r(N \sqcup_\partial (-N))$.

The original definition of the Turaev-Viro invariants in [57] made use of a state sum, where a state is defined as an admissible coloration of the 1-skeleton of a triangulation, and the summand corresponding to a state is the product of contributions for each of the tetrahedra, faces and edges of the triangulation (in particular, the contribution of a tetrahedron is a quantum $6j$-symbol). Roberts shows that this definition is equivalent (up to normalization) to the definition given above, making use of the fusion rule recalled in the previous section. In the present setting one must recall (Proposition 2.3.5) that a standard spine defines by duality (see Section 2.3) a "triangulation" of M with one vertex only.

In [7] we have shown how to describe a diagram of Roberts' framed link $L(\Gamma)$ starting from any o-graph. The most delicate point was to compute the exact framing, and then to show that when one goes on to compute $\langle\!\langle\,.\,\rangle\!\rangle$ the framing is equivalent to the obvious diagrammatic one (this is needed to prove the equality with the Turaev-Viro invariant).

In connection with what just stated, we remark soon one pleasant effect of using branched standard spines and normal o-graphs, rather than just standard spines and o-graphs. Recall that if H is a handlebody obtained by thickening Γ, then the components of $L(\Gamma)$ are just the attaching circles of the discs of $P(\Gamma)$ to ∂H, where the framing is given by taking a regular neighbourhood (a strip) in ∂H. Now, if Γ is a normal o-graph, then these strips are always essentially vertical. Therefore *for a normal o-graph Γ the correct framing of Roberts link $L(\Gamma)$ is given by the obvious diagrammatic framing.*

Refined TV Invariants Let M be a closed oriented 3-manifold. Recall that we have the following bijections:

$$\Theta : H_1(M) \times H_2(M) \;\to\; H^1(M) \times H^1(-M) \cong H^1(M\#(-M))$$
$$(x, y) \;\mapsto\; (x, x + D(y))$$
$$\Phi : \mathrm{Spin}(M) \times H_2(M) \;\to\; \mathrm{Spin}(M\#(-M)) \cong \mathrm{Spin}(M) \times \mathrm{Spin}(-M)$$
$$(s, y) \;\mapsto\; (s, -s + D(y))$$

If $r \equiv 2$ or $r \equiv 0$ modulo 4 we respectively define:

$$\mathrm{TV}^{r(2)}(M, x, y) = \eta \mathrm{RTW}^{r(2)}(M\#(-M), \Theta(x, y))$$
$$\mathrm{TV}^{r(0)}(M, s, y) = \eta \mathrm{RTW}^{r(0)}(M\#(-M), \Phi(s, y)).$$

It follows directly from these definitions that $TV^{r(i)}$ is indeed an invariant for $i = 0, 2$; moreover the properties of the invariants $RTW^{r(i)}$ (similar to those already used to state the Turaev-Walker theorem) imply that:

$$TV^{r(2)}(M, x, y) = RTW^{r(2)}(M, x) \cdot RTW^{r(2)}(-M, x + D(y))$$
$$TV^{r(0)}(M, s, y) = RTW^{r(0)}(M, s) \cdot RTW^{r(0)}(-M, -s + D(y)).$$

Computation of refined TV invariants. We can illustrate now how to compute these refined invariants using the normal o-graphs of the spin calculus described in Theorem 1.4.4. Fix an o-graph Γ representing a closed manifold M, fix a $\mathbb{Z}_2$-colouring of the edges of Γ which defines a spin structure s on M, and fix $y \in H_2(M)$. We know how to construct a framed link $L = L(\Gamma)$ which represents $M \#(-M)$ and therefore can be used in computing $RTW^{r(0)}$. Let us remark again that the framing of L is the obvious diagrammatic one.

Now we have to determine the colours Ω_0 or Ω_1 for the various components of L. Recall that if Γ has $g - 1$ vertices then L consists of the g attaching circles of the discs of $P(\Gamma)$, together with a system of g meridians of the handlebody H obtained as a neighbourhood of the loosened version of Γ. Since the given class $y \in H_2(M)$ comes as a $\mathbb{Z}_2$-linear combination of discs, we just give each attaching curve $\partial \Delta$ the colour corresponding to the coefficient of the disc Δ in y.

We are left to determine the colours we must give to the meridians. In [48] one can find the following result (which we express in our terminology):

Lemma 8.3.1. *Consider one of the meridians μ, and select a curve γ in H which links μ once, and does not link any of the other meridians; γ can be regarded both as a curve in M and as a curve in $\mathbb{R}^3$, and hence there are two spin structures defined along it. Then μ must be given colour Ω_0 or Ω_1 depending on whether these two spin structures agree or not.*

Let us explain how to turn this process into an effective algorithm. We start by selecting a maximal tree Γ_0 in Γ (this will have $g - 2$ edges). By repeatedly applying the operation of adding the coboundary of a vertex (move of Fig. 1.7) we can assume that the $\mathbb{Z}_2$-colouring which defines the spin structure is 0 on Γ_0. Without loss of generality we can also assume that Γ_0 is embedded in the plane, and that each of the other edges is simple and does not meet Γ_0 (except at the ends). Moreover we can construct the link $L(\Gamma)$ by choosing one meridian for each of the edges not in Γ_0, as a loop which encircles the edge. To determine if one such meridian μ should have colour 0 or 1, we proceed as follows:

1. We select the edge $\ell \subset \Gamma \setminus \Gamma_0$ which corresponds to μ.

2. We select the only simple path $\tilde{\ell} \subset \Gamma_0$ which joins the ends of ℓ.

3. Along $\ell \cup \tilde{\ell}$, which can be viewed as a simple piecewise C^1 loop in the plane, we consider the tangent field v used to define the Euler cochain (see Fig. 7.2).

4. We compute the winding number w of v along $\ell \cup \tilde{\ell}$, which is of course well-defined.

5. Now μ should be given the colour Ω_i, where $i \equiv w + c$ modulo 2, and c is the colour of ℓ.

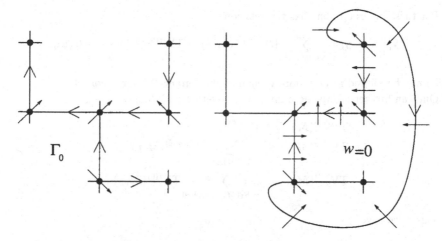

Figure 8.2: Computation of winding number

An example of this process is given in Fig. 8.2. We summarize our construction by:

Proposition 8.3.2. *The procedure just described allows an effective computation of the invariants* $\mathrm{TV}^{r(0)}(M, s, y)$.

For the case $r \equiv 2$ modulo 4 one could describe a similar process to compute the invariants $\mathrm{TV}^{r(2)}(M, x, y)$, but in this case the advantage of dealing with normal o-graphs is just that one does not need to worry about the framing.

Let us remark now that also for the refined skein theory there exist fusion rules analogue to that of Fig. 8.1, only more complicated. Therefore one can in principle express also the refined invariants by means of state sums, but the formulae are rather elaborated. Easier state sums allow to compute certain combinations of the refined invariants, which we will now describe.

Combinations of TV invariants. We will only refer to the case $r \equiv 0$ modulo 4. With some formal variations the definitions could be easily carried over to the case $r \equiv 2$ modulo 4.

Let us first remark that by definition for any closed oriented 3-manifold M we have

$$\mathrm{TV}^r(M) = \sum_{s \in \mathrm{Spin}(M),\, y \in H_2(M)} \mathrm{TV}^{r(0)}(M, s, y).$$

If we fix y we can define another invariant for pairs (M, y)

$$\mathrm{TV}^{r(0)}(M, y) = \sum_{s \in \mathrm{Spin}(M)} \mathrm{TV}^{r(0)}(M, s, y)$$

which was already pointed out in [57]. This invariant can be computed by restricting the state sum which gives the general TV^r to admissible colourings which represent y in the following sense: a colouring can be seen as a linear combination of 2-cells with coefficients in $\{0, 1, \ldots, r-2\}$, and admissibility easily implies that its modulo 2 reduction defines a $\mathbb{Z}_2$-cycle, and therefore represents an element of $H_2(M)$.

For these invariants one has the relation

$$\mathrm{TV}^{r(0)}(M,y) = \sum_{s \in \mathrm{Spin}(M)} \mathrm{RTW}^{r(0)}(M,s) \cdot \mathrm{RTW}^{r(0)}(-M, -s + D(y)),$$

which can be viewed as one more version of the Turaev-Walker theorem.

One can further group these invariants together and define:

$$
\begin{aligned}
\mathrm{TV}_0^{r(0)}(M) &= \mathrm{TV}^{r(0)}(M, 0) \\
\mathrm{TV}_1^{r(0)}(M) &= \sum_{y \in H_2(M), y^3 = 1} \mathrm{TV}^{r(0)}(M, y) \\
\mathrm{TV}_2^{r(0)}(M) &= \sum_{y \in H_2(M), y \neq 0, y^3 = 0} \mathrm{TV}^{r(0)}(M, y).
\end{aligned}
$$

Of course we have

$$\mathrm{TV}^r(M) = \mathrm{TV}_0^{r(0)}(M) + \mathrm{TV}_1^{r(0)}(M) + \mathrm{TV}_2^{r(0)}(M).$$

Since in Chapter 10 we will show how to compute explicitly y^3 in the setting of normal o-graphs, also $\mathrm{TV}_0^{r(0)}(M)$, $\mathrm{TV}_1^{r(0)}(M)$ and $\mathrm{TV}_2^{r(0)}(M)$ can be effectively computed using our technology.

In Section 10.2 we will show that for a spin manifold (M, s) one can define a map $z_s : H_2(M) \to \mathbb{Z}_4$ which lifts the operation $y \mapsto y^3$. Hence we can define for $i \in \mathbb{Z}_4$ the following invariants which again can be effectively computed using the spin calculus:

$$(M, s) \mapsto \sum_{y \in H_2(M), z_s(y) = i} \mathrm{TV}^{r(0)}(M, s, y).$$

8.4   An alternative computation of TV invariants

Consider a closed spin manifold (Q, s). A celebrated theorem due to Milnor (see [27]) states that there exists a framed link L in S^3 such that Q is obtained by Dehn surgery along L and s has image 0 under the bijection

$$\Psi_L : \mathrm{Spin}(Q) \to \mathcal{K}(L)$$

defined earlier in this chapter. In other words this means that the restriction of s to $S^3 \setminus L$ coincides with the restriction of the unique spin structure of S^3, and that s extends to a spin structure on the 4-manifold W_L whose boundary is Q. This also implies that the invariant $\mathrm{RTW}^{r\ (0)}(Q, s)$ can be computed by a formula in which only the colour Ω_0 appears.

We will now give an elementary proof of Milnor's theorem for spin manifolds of the form $(M \# (-M), (s, -s))$. This proof is based on the framed link $L(\Gamma)$ already mentioned in this chapter, obtained from an o-graph Γ of the punctured version of M. The key technical result is the following lemma, of which we outline two proofs.

Lemma 8.4.1. *Let H be a handlebody and let $c_1, \ldots, c_n$ be pairwise disjoint simple curves on ∂H, framed by a regular neighbourhood in ∂H. Then there exists an embedding of H in $\mathbb{R}^3$ such that all the c_i's, viewed as curves in $\mathbb{R}^3$, have even framing.*

Proof of 8.4.1. We start with a direct argument. First of all, assume that H is embedded in $\mathbb{R}^3$ as the standard handlebody of genus g, and choose the standard meridians μ_j and longitudes λ_j for H, with $j = 1, \ldots, g$. Now note that any curve on ∂H, if viewed as an element of $H_1(\partial H)$, can be expressed as $\sum_{j=1}^{g}(\alpha_j \mu_j + \beta_j \lambda_j)$, and the class modulo 2 of the framing of the curve is given by $\sum_{j=1}^{g} \alpha_j \beta_j$. Moreover given two transversal curves and the corresponding expressions $\sum_{j=1}^{g}(\alpha'_j \mu_j + \beta'_j \lambda_j)$ and $\sum_{j=1}^{g}(\alpha''_j \mu_j + \beta''_j \lambda_j)$, then the class modulo 2 of the number of intersection points of the curves is $\sum_{j=1}^{g}(\alpha'_j \beta''_j + \alpha''_j \beta'_j)$.

Now we denote by T_j the Dehn twist along μ_j; we will allow ourselves to modify the embedding by applying combinations $T_1^{x_1} \cdots T_g^{x_g}$. Note that T_j naturally operates on $H_1(\partial H)$, where $T_j(\lambda_j) = \lambda_j + \mu_j$ and T_j is the identity on the other $2g - 1$ generators.

Let $\sum_{j=1}^{g}(\alpha_j^{(k)} \mu_j + \beta_j^{(k)} \lambda_j)$ correspond to c_k. Then we have the condition

(C) $$\sum_{j=1}^{g}(\alpha_j^{(k)} \beta_j^{(l)} + \alpha_j^{(l)} \beta_j^{(k)}) = 0 \qquad k, l = 1, \ldots, n,$$

and we want to show that the $\mathbb{Z}_2$-linear non-homogeneous system

(S) $$\sum_{j=1}^{g}(x_j + \beta_j^{(k)}) \alpha_j^{(k)} = 0 \qquad k = 1, \ldots, n$$

admits a solution $x \in (\mathbb{Z}_2)^n$. Note that here we have used the fundamental property that $z^2 = z$ for all $z \in \mathbb{Z}_2$; the same property is tacitly used several times in the sequel.

We have shown that the initial topological problem reduces to $\mathbb{Z}_2$-linear algebra: (S) should have a solution whenever (C) holds. This is easily shown by induction on g, we just sketch the argument. The case $g = 1$ is easily. For $g > 1$ we note that if $\alpha_j^{(k)} \equiv 0$ any x solves (S), so we assume $\alpha_1^{(1)} = 1$ and we express x_1 as a combination of $x_2, \ldots, x_n$. Using the relations (C) we then show that the system to be solved in $x_2, \ldots, x_n$ is the following analogue of (S)

$$\sum_{j=2}^{g}(x_j + \tilde{\beta}_j^{(k)}) \tilde{\alpha}_j^{(k)} = 0 \qquad k = 2, \ldots, n$$

with $\tilde{\alpha}_j^{(k)} = \alpha_1^{(k)} \alpha_j^{(1)} + \alpha_j^{(k)}$ and $\tilde{\beta}_j^{(k)} = \alpha_1^{(k)} \beta_j^{(1)} + \beta_j^{(k)}$. It is now a routine matter to show that these new constants satisfy the conditions which correspond to (C), and the conclusion follows.

Let us now outline another proof. We first consider the 3-manifold with boundary N obtained by attaching 2-handles to H along $c_1, \ldots, c_n$, and we fix any spin structure s on N. Note that by construction c_i bounds a disc in N in such a way that the framing on c_i extends to the disc; therefore the framing of c_i with respect to s is even. Consider now an embedding of H into $\mathbb{R}^3$. Then H can be naturally endowed with two different spin structures: the restriction $\tilde{s}$ of s and the restriction s_0 of the unique spin structure of $\mathbb{R}^3$. Now, we know that the difference between two spin structures is an element of $H^1(H)$; let h be the difference between $\tilde{s}$ and s_0. We can represent h as a $\mathbb{Z}_2$-linear combination of the 1-handles attached to the 3-ball to obtain H. Now we modify the embedding of H into $\mathbb{R}^3$ by applying a Dehn twist to each of the 1-handles having coefficient 1 in the expression of h. If we denote by s_1 the spin structure on H induced by this new embedding in $\mathbb{R}^3$, then by definition we have $s_1 = \tilde{s}$, and the conclusion follows. $\boxed{8.4.1}$

Given a framed link L in S^3, we will write $f_2(L) = 0$ if all the components of L have even framing. The previous lemma readily implies the following proposition; for the definition of closed normal o-graph refer to Chapter 1.

Proposition 8.4.2. *Let Γ be a closed normal o-graph of a closed 3-manifold M. Then, up to adding some curls to the edges of Γ, for the framed link $L(\Gamma)$ of Definition 2.3.2 we have $f_2(L(\Gamma)) = 0$.*

To be completely formal in the previous proposition one would have to give a new definition of o-graph, omitting the curl-move from the moves which generate the equivalence relation. We avoid to do this explicitly, leaving the details to the reader. Compare also with the ambiguity in the definition of $L(\Gamma)$ already pointed out in Section 2.3.

To show that the previous proposition implies Milnor's theorem for spin manifolds of the form $(M\#(-M), (s, -s))$ we have now some extra work.

Lemma 8.4.3. *If Γ is as above and $f_2(\Gamma) = 0$ then Γ defines a spin structure $s(\Gamma)$ on M which coincides with the unique structure of S^3 in a neighbourhood of Γ.*

Proof of 8.4.3. Recall that a spin structure can be viewed as a trivialization of the tangent bundle on the 1-skeleton which extends to a trivialization on the 2-skeleton, up to homotopy on the 1-skeleton. Now, Γ itself can be taken as the 1-skeleton of a cellularization of M. Then we only have to note that the evenness of the framing guarantees that the trivialization induced on Γ by the embedding in S^3 extends to the 2-skeleton. $\boxed{8.4.3}$

Now recall that the elements of $H^1(M)$ can be expressed as $\mathbb{Z}_2$-linear combinations of edges of Γ. If $h \in H^1(M)$ let $h(\Gamma)$ be obtained by adding a curl on each of the edges of Γ which have coefficient 1 in h. We leave the following to the reader.

Lemma 8.4.4. *If $h \in H^1(M)$ and $f_2(L(\Gamma)) = 0$ then $f_2(L(h(\Gamma))) = 0$. Moreover $s(h(\Gamma)) = s(\Gamma) + h$, where the $+$ sign refers to the natural structure on $\mathrm{Spin}(M)$ of affine space over $H^1(M)$.*

The previous lemma implies in particular that one could base on closed normal o-graphs Γ with $f_2(\Gamma) = 0$ a representation of closed spin 3-manifolds. The following is another result which we leave to the reader; as a corollary to it we state the special case of Milnor's theorem which we have proved in this context.

Lemma 8.4.5. *If $f_2(\Gamma) = 0$ then the spin structure on $S^3 \setminus L(\Gamma)$ induced by the unique spin structure of S^3 extends to the spin structure $(s(\Gamma), -s(\Gamma))$ on the manifold $M\#(-M)$ obtained by surgery along $L(\Gamma)$.*

Corollary 8.4.6. *Given a spin structure s on a closed oriented 3-manifold M there exists a closed normal o-graph Γ of M such that $f_2(L(\Gamma)) = 0$ and $L(\Gamma)$ induces on $M\#(-M)$ the spin structure $(s, -s)$.*

Going back to the computation of the spin-refined Turaev-Viro invariants, let us note now that if Γ is as in the previous corollary then $TV^r{}^{(0)}(M, (s, 0))$ is obtained by giving all the components of $L(\Gamma)$ the colour Ω_0. Similarly for $TV^r{}^{(0)}(M, (s, y))$ one has to colour by Ω_1 exactly the components of $L(\Gamma)$ which correspond to discs having coefficient 1 in y.

We conclude by putting forward a question. We know that, by means of appropriate fusion rules, one can express the invariant $\mathrm{TV}^{r\,(0)}(M,(s,0))$ by means of a state sum. We have just shown that the invariant, which a priori involves the computation of the bracket on a framed link with colours Ω_0 and Ω_1, can actually be computed with the colour Ω_0 only. This fact might have qualitative effects on the state sum giving the invariant, which it would be interesting to investigate.

Chapter 9

Problems and perspectives

In this chapter we will briefly illustrate some problems which arise in connection with the theory of branched standard spines developed in this monograph, and some fields where this theory might have applications. This chapter is written in a less formal and detailed style than the rest of this work. In particular we will omit most of the proofs of the facts we state, and also some definitions, addressing the reader to the existing literature.

The three sections of this chapter correspond respectively to the distinction of the questions we rise into the following three categories: those internal to the theory of branched spines, those connected with the theory of (quantum) invariants, and those regarding extra geometric structures, namely foliations and contact structures.

9.1 Internal questions

Let us recall from the Matveev-Piergallini theory (Chapter 2) that two standard spines with at least two vertices of the same manifold with boundary are connected by a finite combination of the MP move and inverses of it. Moreover in Section 3.4 we have shown that the MP move can be interpreted as a desingularization move (any fundamental 2-chain, viewed as a singular branching, can by desingularized via MP's to a genuine branching). By analogy with the situation in algebraic geometry we formulate the following:

Question 9.1.1. *Given standard spines P_1, P_2 of the same manifold M with boundary, is it possible to find a standard spine P of M for which there exist two sequences of positive MP moves which start from P_1 and P_2 respectively and both lead to P?*

Another motivation for the previous question is given by the analogy with Heegaard splittings of closed manifolds. Note that a standard spine P of a closed manifold M defines a Heegaard splitting of M, and that an MP move on P defines an elementary stabilization of the Heegaard splitting. Now, it is well-known that for any two Heegaard splittings of M there is another one which is a stabilization of both. This fact itself is of course not sufficient to answer in the affirmative to Question 9.1.1, because it is not true that every elementary stabilization is carried by an MP move.

◇   ◇   ◇

Recall from Section 3.1 that we have denoted by $\mathcal{F}_2(P)$ the set of fundamental 2-chains for an (oriented) standard spine P (i.e. essentially the orientations of the discs of P; recall also that the set $\mathcal{B}(P)$ of branchings is a subset of $\mathcal{F}_2(P)$). Moreover in Section 3.4 we have noticed that, if $P \to P'$ is a positive MP move, a 2-to-1 map $\mathcal{F}_2(P') \to \mathcal{F}_2(P)$ is naturally defined.

Let us consider the set $\mathcal{F}$ of all pairs (P, γ) where P is a standard spine with at least two vertices and $\gamma \in \mathcal{F}_2(P)$. On this set we consider the moves $(P, \gamma) \to (P', \gamma')$ where $P \to P'$ is a positive MP move and γ' is one of the two elements of $\mathcal{F}_2(P')$ which correspond to γ. Let $\sim$ be the equivalence relation on $\mathcal{F}$ generated by these moves. By the Matveev-Piergallini theory we know that if (P, γ) and (P', γ') are equivalent under $\sim$ then P and P' define the same manifold. We do not know if the converse is true:

Question 9.1.2. *Is the coset space $\mathcal{F}/_\sim$ naturally identified to the set of (compact, oriented, connected) 3-manifolds?*

Note that the elements of $\mathcal{F}_2(P)$ do not have an obvious geometric interpretation (while branchings do have this interpretation, and we have based on it all the constructions of this monograph). This makes Question 9.1.2 not completely trivial from the combinatorial point of view, and also only mildly interesting from the geometric point of view.

The following is a geometrically more significant version of the previous question. Let us denote by $\mathcal{B}$ the subset of $\mathcal{F}$ given by pairs (P, γ) where γ is a branching on P. On $\mathcal{B}$ we have the moves of Figures 1.4 and 1.5 (which generate the flow-preserving calculus, in the sense of Section 4.3) and the move of Figure 3.22, the effect of which was analyzed in Section 4.6. Let these moves generate the relation $\sim$ on $\mathcal{B}$.

Question 9.1.3. *Is the coset space $\mathcal{B}/_\sim$ naturally identified to the set of (compact, oriented, connected) 3-manifolds?*

An affirmative answer to this question would imply an affirmative answer to Question 9.1.2, via desingularization (Theorem 3.4.9).

Concerning Question 9.1.3, recall from Chapter 4 that every element of $\mathcal{B}$ defines a traversing flow on the corresponding manifold M, and that the move of Figure 3.22 allows to transform into each other any two flows viewed as vector fields up to homotopy on M. However the moves of Figures 1.4 and 1.5 translate an equivalence relation which is more refined than just homotopy (see Section 4.1). Therefore one way to attack Question 9.1.3 could be to carefully analyze the difference between these two relations.

We can also specialize the previous question to the case of closed manifolds, by restricting to the set $\mathcal{B}_c$ of pairs (P, γ) where P defines a manifold N bounded by S^2 (but without restrictions on the bicoloration induced by γ). We consider on $\mathcal{B}_c \subset \mathcal{B}$ the restricted equivalence relation.

Question 9.1.4. *Is the coset space $\mathcal{B}_c/_\sim$ naturally identified to the set of (closed, oriented, connected) 3-manifolds?*

We have noticed in Section 4.2 that it is always possible to extend the flow carried by γ to the closed manifold $M = \widehat{N}$. A positive answer to Question 9.1.4 would lead immediately to a new combinatorial realization of closed manifolds, on which a realization of spin manifolds could probably be based. These realizations would be combinatorially simpler than those illustrated in this book, because the Pontrjagin move would not have to be taken into account.

◇   ◇   ◇

We conclude this section with the following very natural problem. Recall that the various calculi established in this work (for closed, combed, framed and spin manifolds) are all generated by finite but rather huge sets of moves.

Question 9.1.5. *Is it possible to find smaller generating sets of moves for the various calculi? Are the moves of the present calculi independent?*

9.2   Questions on invariants

Quantum invariants. We have illustrated in Chapter 8 how (suitably decorated) branched spines could be used to implement the construction of (spin-refined) Turaev-Viro invariants. To do this we have only exploited our combinatorial realization of spin 3-manifolds. However branched spines also allow to encode framings on closed manifolds. Now, according to Witten's interpretation, the Reshetikhin-Turaev-Witten invariants are invariants of bi-framed manifolds: roughly speaking, if f is an ordinary framing on M, we can consider the trivialization (f, f) of $TM \oplus TM$; moreover $TM \oplus TM$ can be endowed with its natural spin structure. Hence (f, f) can be regarded up to homotopy on $TM \oplus TM$. Now, one has invariants $\mathrm{RTW}(M, [(f, f)])$, and the current interpretation of the model of absolute RTW invariant recalled in Chapter 8 is then given through the canonical bi-framing, in the sense of Atiyah [1].

Question 9.2.1. *Is it possible, using the combinatorial presentation of framed manifolds, to determine the framing which gives rise to the canonical bi-framing?*

Question 9.2.2. *Is it possible to compute the invariants $\mathrm{RTW}(M, [(f, f)])$ directly from a combinatorial presentation of the closed framed manifold (M, f)?*

Let us remark now that the invariants of Turaev-Viro type are all uneffected by the natural involutions associated to the change of orientation (e.g. for the absolute invariant we have $\mathrm{TV}(M) = \mathrm{TV}(-M)$, and similarly for the spin-refined versions). We also note that these involutions very simply translate into involutions on the graphs which we employ in our combinatorial realizations. This motivates the following:

Question 9.2.3. *Is it possible to combine the technology of quantum invariants and our combinatorial realizations to produce invariants which are sensitive to the change of orientation?*

The previous questions concerning invariants would be solved if the following purely topological question had a solution. Recall that according to [56] a *shadow* is a combinatorial encoding of a 4-manifold, where the encoding is supported by a simple 2-polyhedron.

Question 9.2.4. *Does there exist an effective procedure to construct, starting from a branched standard spine of the framed calculus representing a 3-manifold M, a surgery or shadow presentation of a 4-manifold W with $\partial W = M$?*

$$\diamond \qquad \diamond \qquad \diamond$$

Kupenberg invariants. G. Kupenberg [33] has developed an alternative approach to the construction of invariants of 3-manifolds, starting from (arbitrary) Hopf algebras. In their most general version these are invariants of framed (or, at least, combed) manifolds (moreover they are potentially sensitive to the change of orientation, in the sense explained above). Therefore Kupenberg's theory is a very suitable field to test the efficiency of our combinatorial presentations; this is why we will recall it here with some detail. Following [33] we will employ for Hopf algebras the arrow notation instead of the more traditional index notation.

For a (complex) vector space V we will denote a tensor $T \in V_1 \otimes \cdots \otimes V_j$, with $V_i \in \{V, V^*\}$, by the same letter T together with an outgoing arrow for each V-factor, and an incoming arrow for each V^*-factor. For instance $v\to$ denotes a vector and $\to w$ denotes a covector. A permutation of the arrows corresponds to a permutation of the factors: for instance the symbolic relation

$$\overset{\Large\searrow}{\to} g = \times\!\!\!\overset{\to}{\to} g$$

translates the fact that g is a symmetric bilinar form on V. The tensor product of two tensors is represented by juxtaposition; for instance $\begin{smallmatrix} v_1 \to \\ v_2 \to \end{smallmatrix}$ corresponds to $v_1 \otimes v_2$ in $V \otimes V$. The symbol $v\to L\to$ will denote the vector $L(v) \in V$; note that this vector is also represented by $L(v)\to$, which means that the composition rules for the objects are the natural ones. Finally the arrow $\to$ will denote the identity map on V and (if V is finite-dimensional) we will define

$$\overset{L}{\underset{\frown}{\frown}} = \mathrm{tr}(L), \quad \text{in particular} \quad \bigcirc = \mathrm{tr}(\mathrm{Id}_V) = \dim(V).$$

Now let H be a finite-dimensional Hopf algebra. The axioms of H can be summarized as follows: there exist tensors $\overset{\to}{\to}M\to$ (multiplication), $\to\!\Delta\!\overset{\to}{\underset{\to}{}}$ (comultiplication), $i\to$ (unit), $\to\epsilon$ (counit), $\to S\to$ (antipode) such that the following relations are verified:

$$\overset{\diagup}{\underset{\to}{\searrow}}M\overset{\diagup}{\underset{}{}}M\to = \overset{\to M\searrow}{\underset{\diagup}{\searrow}}M\to \qquad \text{(associativity)},$$

$$\to\!\Delta\!\underset{\Delta\overset{\prec}{\prec}}{\overset{\diagup}{\searrow}} = \to\!\Delta\!\overset{\Delta\overset{\prec}{\prec}}{\underset{\searrow}{\diagup}} \qquad \text{(coassociativity)},$$

$$_i\!\overset{\to}{\to}M\to = {}^i\!\overset{\to}{\to}M\to = \to\!\Delta\!\overset{}{\underset{\epsilon}{\prec}} = \to\!\Delta\!\overset{\epsilon}{\prec} = \to, \qquad i\to\epsilon = 1 \quad \text{(unit and counit)},$$

$$\succ M \to \Delta \prec \; = \; \times{\to \Delta \to M \to \atop \to \Delta \to M \to} \qquad \text{(bialgebra)},$$

$$\to \Delta \prec^{S} \succ M \to \; = \; \to \Delta \prec_{S} \succ M \to \; = \; \to \epsilon \cdot i \to \qquad \text{(antipode)},$$

(in the right-most member of the last relation we are using $\cdot$ instead of juxtaposition, because in this case the meaning is exactly complex multiplication: in other words $\to \epsilon \cdot i \to$ is the linear map $(\to \epsilon) \otimes (i \to)$ which takes a vector v to $\varepsilon(v) \cdot (i \to)$).

We also define:

$$k\left\{ \vdots \succ M \to \; = \; \underbrace{\succ M \succ M \to \cdots \succ M \to}_{k-1} , \qquad k\left\{ \vdots \succ \Delta \leftarrow \; = \; \underbrace{\to \Delta \to \Delta \leftarrow \cdots \to \Delta \leftarrow}_{k-1} . \right.\right.$$

Two covectors $\to \mu_R$ and $\to \mu_L$ will be called respectively a right and left integral if

$$\to \Delta \prec^{\mu_R} = \; \to \mu_R \cdot i \to , \qquad \to \Delta \prec_{\mu_L} = \; \to \mu_L \cdot i \to .$$

Two vectors $e_R \to$ and $e_L \to$ will be called respectively a right and left cointegral if

$$e_R \succ M \to = \; \to \epsilon \cdot e_R \to , \qquad e_L \succ M \to = \; \to \epsilon \cdot e_L \to .$$

It is possible to check that every Hopf algebra admits essentially unique integrals and cointegrals. Moreover if we define

$$\to P_R \to \; = \; \to M \prec^{S}_{S} \succ \Delta \to , \qquad \to P_L \to \; = \; \to M \prec^{S}_{S} \succ \Delta \to$$

then we have $\operatorname{tr}(P_R) = \operatorname{tr}(P_L) = 1$ and $\operatorname{tr}(P_R^2) = \operatorname{tr}(P_L^2) = \sigma = \pm 1$, and if $\sigma = 1$ then

$$\to \mu_R \cdot e_R \to \; = \; \to P_R \to , \qquad \to \mu_L \cdot e_L \to \; = \; \to P_L \to .$$

The theory is more complicated in the case $\sigma = -1$, because certain "sign-orderings" must be introduced (and affect also the definition of the invariants). We do not want to get into technical details, therefore we will always assume in the sequel that $\sigma = 1$.

Having fixed integrals and cointegrals let us define:

$$a \to \; = \; e_R \to \Delta \prec_{\mu_R} \qquad \qquad \to \alpha \; = \; e_R \succ M \to \mu_R.$$

It turns out that $a \to$ is a group-like element in H and $\to \alpha$ is a group-like element in the dual algebra H^*. We recall that g is group-like in H if

$$g \to \Delta \prec \; = \; {g \to \atop g \to} , \qquad \qquad g \to \epsilon = 1$$

and moreover, having set $g^{-1} = S(g)$, for all $n \in \mathbf{Z}$, $|n| \geq 2$ the following notation is unambiguously defined:

$$g^n = \; n\left\{ {g \atop \vdots \atop g} \succ M \to . \right.$$

Finally we define for all $n \in \mathbf{Z}$

$$\to \mu_n = \; _{a^n} \succ M \to \mu_R \qquad e_n \to \; = \; e_R \to \Delta \prec_{\alpha^n} \qquad \to T \to \; = \; \to S^{-2} \to \Delta \prec^{\alpha^{-1}}_{\alpha} .$$

We recall that the definition of $\to T \to$ is related to the validity of the following important relation:

$$\begin{array}{c} a \searrow \\ {} \rightrightarrows M \to \Delta \begin{array}{c} \nearrow \alpha^{-1} \\ \leftrightarrows \\ \searrow \alpha \end{array} = \to S^4 \to . \\ a^{-1} \nearrow \end{array}$$

The Hopf algebra H is said to be *balanced* if $\to T \to \; = \; \to$.

We are now ready to describe the definition of Kupenberg invariants. We will use our terminology of branched standard spines (which induce Heegaard splittings, used by Kupenberg). Let P be a branched standard spine of a manifold N bounded by S^2 with trivial bicoloration on the boundary, and consider a $\mathbb{Z}_2$-colouring of the edges of P which encodes a framing of $M = \widehat{N}$, according to Theorem 1.4.3 (we will briefly express this by saying that P is a coloured branched standard spine which represents a framed closed 3-manifold).

The idea of the invariant is to attach to the various discs and edges of P certain tensors depending on parameters, in such a way that the various arrows match and globally give rise to a tensor without arrows (i.e. a scalar). Then one must show that a choice of the parameters is possible such that the result does not depend on P but only on the framed manifold M.

Recall that if P has $g - 1$ vertices then it defines a Heegaard splitting of M of genus g, obtained by taking a regular neighbourhood U of $S(P)$, and the attaching curves on ∂U of the discs of P. Let us select g edges of $S(P)$ such that their complement in $S(P)$ is connected (these will correspond to a system of meridians of U, as in the construction explained in Section 2.3). The chosen set of edges will be denoted by $\mathcal{C}$. In the figures an edge in $\mathcal{C}$ will be marked by a tick.

Let D be a disc of P. We will use an abstract version of D, subdividing its boundary circle into the various segments which get glued to the edges of P. We associate to D a Δ-type tensor, with one arrow for each edge in $\mathcal{C}$ on ∂D pointing from Δ towards the corresponding edge, and one arrow pointing from a certain $e_k = e_k(D)$ to Δ. Since the cyclic order of the arrows around Δ makes difference, in this definition we need to arbitrarily fix a basepoint (denoted by $*$) around the Δ, from which the arrow $e_k \to \Delta$ points. See Figure 9.1. Here k is a parameter which we leave undetermined for the time being.

Now to each of the edges in $\mathcal{C}$ we attach an M-tensor with one incoming arrow from each of the surrounding discs (note that each of these arrows comes from some Δ) and an outgoing arrow which points from M to some μ_h, where h is a parameter which depends on the edge. As above to specify the cyclic order of arrows around M we fix some basepoint $*$ (see Figure 9.2).

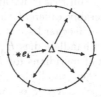

Figure 9.1: The Δ-tensor associated to a disc

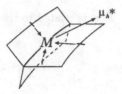

Figure 9.2: The M-tensor associated to an edge

To conclude, we obtain a closed tensor by inserting a tensor $\to T^m S^n \to$ on each of the ticked arcs on the boundaries of the discs (so that locally we see $\Delta \to T^m S^n \to M$); again m and n are parameters which we do not prescribe.

The main result in [33] can be stated in the following (deliberately vague) way:

Theorem 9.2.5.   *1. Given an arbitrary Hopf algebra H, there exists a procedure with the following input:*

- *a coloured branched standard spine P of a some framed closed 3-manifold,*
- *a choice of the basepoints $*$ on P in the above-described process,*

and the following output:

- *a choice of the parameters k, h, m, n in the process,*

with the property that the number associated to P via the process, with the given choices of $*$'s and parameters, depends only on the framed manifold represented by P. Therefore H allows to define an invariant of framed closed manifolds.

2. If H is balanced a similar procedure exists which applies to the spines of the combed calculus, leading to an invariant of combed closed manifolds.

Having expressed this result in purely existencial terms, we are naturally lead to the following:

Question 9.2.6. *It is possible to make effective the procedures whose existence is guaranteed by Theorem 9.2.5?*

An answer in the affirmative could also lead to an answer to the following rather interesting problem:

Question 9.2.7. *Do there exist explicit formulae to compute the Kupenberg invariants in terms only the combinatorial data of a presentation of a framed or combed 3-manifold?*

We will explain now why a solution of these problems is not so immediate, starting from the work of Kupenberg. The main reason is that our encoding of framings and combings on 3-manifolds is considerably different than Kupenberg's. Let us briefly sketch his method. Given a genus g Heegaard splitting of a closed oriented manifold M, with a surface F separating two handlebodies H_+ and H_-, let $C_+ = \{c_+^1, \cdots, c_+^g\}$ and $C_- = \{c_-^1, \cdots, c_-^g\}$ be systems of meridians for H_+ and H_- respectively, in generic position with respect to each other (recall that (F, C_+, C_-) is sometimes called a Heegaard diagram of M). We will say a field ξ of tangent vectors to F is *adapted* to (F, C_+, C_-) if:

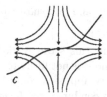

Figure 9.3: Vector field adapted to a system of curves

1. ξ has exactly $1 + 2g$ isolated singularities, one of index 2 and $2g$ of index -1;

2. every curve in $C_+ \cup C_-$ contains exactly one of the singularities of index -1, and does not contain the singularity of index 2;

3. for $c \in C_+ \cup C_-$, at the singularity of ξ contained in c, the curve c is tangent to the two trajectories which originate from the singularity (Figure 9.3).

Kupenberg shows the following:

1. every vector field adapted to a system (F, C_+, C_-) can be thickened to a combing on M;

2. every combing (up to homotopy) is obtained as a thickening of some vector field adapted to a fixed system (F, C_+, C_-);

3. the "calculus" of Heegaard diagrams can be refined to a calculus for adapted vector fields which completely translates the homotopy of combings, thus leading to a presentation of combed manifolds;

4. by restricting to the case of combings with zero Euler class and suitably refining the objects and the calculus, one gets a presentation of framed manifolds;

5. given an adapted vector field one can use the singular points of index -1 as basepoints $*$ in a process formally analogous to what we sketched before Theorem 9.2.5, and hence determine the corresponding parameters k, h, m, n and establish the invariants.

Let us remark that Kupenberg's presentations of combed and framed manifolds are not combinatorial realizations in the sense we have stated in Chapter 1: in particular no combinatorial encoding of adapted fields is given in [33]. Moreover the definition of the invariants is somewhat existential, because it uses various qualitative properties of the field, in particular for the determination of the parameters k, h, m, n, which is based on the variation of certain angles (with respect to a fixed conformal structure on F) depending on the vector field ξ.

However the proof of invariance is indeed analogous to the models of invariants based on combinatorial realizations, because it consists in the case-by-case analysis of the effect of certain elementary moves.

We conclude by indicating a possible strategy towards an answer to Question 9.2.6:

1. Starting with a combing v carried by some branched standard spine, consider the Heegaard splitting also carried by the spine, and find a vector field ξ adapted to it in such a way that ξ defines a combing homotopic to v.

2. Compute the invariant using ξ and translate the result in terms of the initial combinatorial data.

We note however that on one hand this strategy does not appear to be so easily implementable, and on the other hand it is not likely to shade new light on the meaning of the invariants. A more satisfactory approach could be to postulate the existence of the invariants and on this basis find combinatorial formulae which necessarily have to be verified.

9.3   Questions on geometric structures

Our starting point is that on a closed oriented 3-manifold every homotopy class of co-oriented plane fields can be represented both by a foliation [59] and by a contact structure [36], [35], [22], [15]. Moreover homotopy classes of co-oriented plane fields are combinatorially realized by the calculus for combings, so it is very natural to investigate the relations of branched spines with the results just mentioned.

Foliations.   Let P be a spine of the combing calculus, i.e. a branched standard spine of a manifold N with $\partial N = S^2$ and trivial bicoloration on S^2. We set $M = \widehat{N}$. Let us denote by D_i the discs of P and assume that there exists a real cellular 2-cycle $y = \sum_i a_i D_i$ where all the a_i's are strictly positive. Such a y will be called a *positive weight system* (see e.g. [42]). It turns out that to such a y we can associate a foliation of N as follows: we denote by $\pi : N \to P$ the projection, for each i we identify $\pi^{-1}(\text{int}(D_i))$ with $[0, a_i] \times \text{int}(D_i)$ and we foliate it with leaves $\{t\} \times \text{int}(D_i)$; then we glue together the various leaves along the singular set, as illustrated on a cross-section in Figure 9.4; it is not hard to check that indeed this leaf-gluing can be coherently carried out at vertices. Note that the foliation is actually well-defined only on the interior of N, or more generally on $N \setminus E$, where E is the equator of the boundary sphere S^2 along which the flow carried by P is tangent to the sphere. Moreover $S^2 \setminus E$ is the union of two leaves homeomorphic to open discs. More precisely, in a collar of ∂N, the foliation is obtained as suggested in the left-hand side of Figure 9.5; in the right-hand side of the same figure we also show how to modify and extend the foliation on the interior of N to a foliation $\mathcal{F}_y$ on M (recall that y is the positive weight system which was the starting point of the construction). Of course $\mathcal{F}_y$ is homotopic to the plane field on M carried by P. This foliation has other interesting features, among which:

1. It is a taut foliation, i.e. there exists an embedded circle which is transverse to the leaves and intersects all of them; according to [41] and [49] this implies that

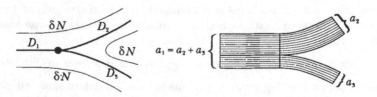

Figure 9.4: Gluing the leaves

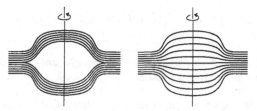

Figure 9.5: Leaves in the neighbourhood of the boundary

M is $S^2 \times S^1$ or an irreducible manifold; (to prove that $\mathcal{F}_y$ is taut, choose a point p_i inside each of the D_i's, take the segments $[0, a_i] \times \{p_i\} \subset \pi^{-1}(\text{int}(D_i))$ and connect these segments with care in the ball $M \setminus N$);

2. There exists a measure transverse to the foliation; since the foliation is co-oriented this implies [54] that M fibres over S^1; (the construction of the transverse measure is again rather elementary).

Therefore, the property for a spine P that a positive weight system should exist yields strong limitations on the topology of the manifold encoded P. Moreover it is easily checked that this property is not invariant under the combing calculus (consider for instance the move of Figure 1.4). Therefore, even if on any given P the existence can be checked in a rather elementary way, the following problem remains:

Question 9.3.1. *Given P, can one decide if there exists some other spine which is equivalent to P under the combing calculus and carries a positive weight system?*

This question could be asked with or without the a priori assumption that the manifold encoded by P is irriducible and fibres over S^1.

◇   ◇   ◇

As we have seen above a positive weight system on a spine P of the combing calculus allows to define a foliation of the closed manifold M associated to P, and this foliation has several interesting properties. We single out some of these properties (omitting the existence of a transverse measure) to give the following definition: a foliation $\mathcal{F}$ of M is said to be *carried* by P if:

1. the open ball $M \setminus N$ can be identified to $(0, 1) \times \text{int}(D^2)$ so that the leaves of the induced foliation are $\{t\} \times \text{int}(D^2)$;

2. on N all the leaves are transversal to the fibres of the projection $\pi : N \to P$;

3. if D is a disc in P then $\pi^{-1}(\text{int}(D))$ can be identified with $[0, 1] \times \text{int}(D)$ with leaves $\{t\} \times \text{int}(D)$.

As above any such $\mathcal{F}$ is taut, so M must be $S^2 \times S^1$ or irreducible. Note that the property of carrying a foliation is probably not invariant under the moves of the combing calculus. The following are very natural questions to ask concerning this notion (as above, one may assume or not to know a priori that M is irreducible):

Question 9.3.2. *Given P, can one decide if P carries a foliation?*

Question 9.3.3. *Given a co-oriented taut foliation $\mathcal{F}$ of M, does there exist a P which carries $\mathcal{F}$?*

Question 9.3.4. *Under which (necessary and/or sufficient) conditions does a manifold admit a spine which carries a foliation?*

Concerning the last question, it could follow from Gabai's work [18] that if M is irreducible and $H_2(M;\mathbb{Z}) \neq 0$ then there exists a spine which carries a foliation; moreover given $0 \neq y \in H_2(M;\mathbb{Z})$ one should be able to find the spine so that the foliation contains a closed leaf which represents y. What seems to be easy to get directly from [18] is a *simple* branched spine, and then one should try to transform it into a standard one (which is not obviously possible, because the property of carrying a foliation does not seem to be invariant under the sliding moves). This motivates the following:

Question 9.3.5. *Given P such that the corresponding M is irreducible and $H_2(M;\mathbb{Z})$ is non-trivial, and given $0 \neq y \in H_2(M;\mathbb{Z})$, is it possible to describe a sequence of moves of the closed calculus (including the Pontrjagin move) which leads from P to a P' which carries $\mathcal{F}$ so that $\mathcal{F}$ contains a closed leaf representing y?*

◇   ◇   ◇

Let us assume now that a foliation $\mathcal{F}$ on M is carried by a certain branched spine P. We want to analyze how far $\mathcal{F}$ is from being carried by a "normalized" positive weight system. To be precise we note that in a neighbourhood of a vertex of index -1 the foliation can be identified to that carried by the local positive weight system shown on the left-hand side of Figure 9.6, and a similar fact holds for vertices of index 1; moreover the foliation along an edge can be identified to that carried by the local positive weight system shown on the right-hand side of Figure 9.6. In other words one can imagine to have foliated building blocks which must be glued together to get $\mathcal{F}$. Of course the gluings must have the property that the holonomy along the boundary of any disc is trivial (because we want $\mathcal{F}$ to be the trivial product foliation over the disc). Moreover the properties of $\mathcal{F}$ depend on these gluings.

Question 9.3.6. *What is the maximal regularity which one can always impose on the gluings arising in the above construction?*

Note that if $\mathcal{F}$ is carried by a positive weight system then the corresponding gluings are affine maps obtained by rescaling linear isometries. Moreover if the gluings are affine and certain global rescaling compatibility conditions hold, then $\mathcal{F}$ is carried by a positive weight system. As we have seen this implies strong topological limitations on

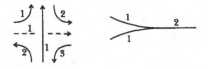

Figure 9.6: Normalizing a foliation

M. Therefore the previous question could be rephrased as follows: how can we weaken the condition that the gluings should be affine to be able to encode all foliations carried by spines?

A weaker version of previous question is obtained by taking into account only spines with a certain limitation on the number of vertices. In this case one might conjecture that the gluings can be realized in some class of "computable" functions (as for instance the semi-algebraic ones, see [9]). Moreover, an optimal solution to the weaker problem might also lead, via asymptotic estimates, to a general solution.

◇   ◇   ◇

Contact structures. A very useful tool in studying contact structures is the characteristic foliation induced on surfaces embedded in the 3-manifold (see [22], [15]). It is therefore tempting to use the characteristic foliation induced by a contact structure ξ on a branched standard spine P, in particular in case ξ is homotopic to the plane field orthogonal to the combing carried by P.

Question 9.3.7. *Is there a notion of a branched standard spine of the combing calculus carrying a contact structure?*

The ingredients of such a notions should be: (1) the construction for any spine P of a contact structure ξ_P on the neighbourhood of P, essentially determined by the corresponding characteristic foliation on P; (2) an extension, as canonical as possible, to the complementary 3-ball. Here are some hints which may be useful. Recall that a contact structure is overtwisted if there exists an embedded disc on which the characteristic foliation has a closed orbit and one singular point inside the orbit, and it is tight otherwise.

1. Construct ξ_P so that the characteristic foliation induced on P is defined by the singular vector field on P used in Section 7.1 to define the Euler cochain.

2. In case P is a spine of the framing calculus, use instead the non-singular vector field on P given by the second vector of the framing. In this case one might also conjecture that the extension of the complementary ball can be taken as the standard tight contact structure on the unit ball of $\mathbb{R}^3$.

3. In the general case for P, try to construct ξ_P so that on one hand the extension can be obtained as in the previous point, and on the other hand the characteristic foliation on P is still determined by the combinatorial data. In this case it would be interesting also to determine how coarsely the geometry of the structure (in particular, its tightness) is determined by the characteristic foliation.

4. Impose that the resulting contact structure should be overtwisted. If this were possible, according to a recent result of Eliashberg [15], it would follow that the combing calculus provides a combinatorial realization of homotopy classes of overtwisted contact structures. Moreover, since two homotopic overtwisted structures are isotopic, we would be able to provide a preferred representative for each homotopy class.

◇   ◇   ◇

Assume a P as above carries a foliation $\mathcal{F}$. Eliashberg and Thurston in a recent work [16] claim that $\mathcal{F}$ can be approximated by a contact structure ξ, which must be tight because $\mathcal{F}$ is taut.

Question 9.3.8. *Can such a ξ (and its characteristic foliation on P) be explicitly described? Can one compare two different such ξ's by comparing their characteristic foliations on P?*

◇   ◇   ◇

Assume that ξ is either a tight contact structure or a taut foliation on M. Then we have the following generalized Bennequin inequalities (see respectively [2]-[14] and [53]) concerning the Euler class of the underlying plane distribution: if F is a closed oriented surface embedded in M then

$$\begin{cases} \mathcal{E}(\xi)[F] = 0 & \text{if } F = S^2 \\ |\mathcal{E}(\xi)[F]| \leq -\chi(F) & \text{otherwise.} \end{cases}$$

Moreover if F has a boundary transversal to ξ then $\mathcal{E}(\xi)[F]$ can be defined (with suitable orientation conventions) in such a way that

$$\mathcal{E}(\xi)[F] \leq -\chi(F).$$

Question 9.3.9. *Given a spine P of the combed calculus, can we decide if the corresponding Bennequin inequalities are verified for the plane field defined by P?*

Recall that a corollary of the Bennequin inequalities ensures that only finitely many elements of $H^2(M; \mathbb{Z})$ can be the Euler class of a tight contact structure or a taut foliation. In other words there is only a finite number of combings on M, viewed up to Hopf number, which can be associated to a tight contact structure or a taut foliation.

Question 9.3.10. *Given P, describe moves of the closed calculus which allow to obtain from P at least one spine representing each of the combings viewed up to Hopf number which verify the Bennequin inequalities.*

Chapter 10

Homology and cohomology computations

This chapter contains computations of homology invariants of 3-manifolds presented by means of our calculi. The first section contains the first elementary homology computations, and its results are used in various parts of this monograph. In the second section we compute some less trivial homology invariants, which have also been referred to in Chapter 8. In the third section we provide a recipe to determine whether a framed knot in a spin manifold is even or odd.

10.1   Homology, cohomology and duality

We start with some elementary remarks concerning the homology and cohomology of branched standard spines, represented by normal o-graphs.

Let P be an (oriented) standard spine with a fixed (oriented) branching, and let Γ be the normal o-graph of P. If we agree to give all the vertices of P a positive sign, then using the branching we have that all the cells of P have a canonical orientation, and therefore to compute the cellular homology and cohomology of P with any coefficient group we only need to compute the boundary operator ∂. The action of ∂ on cells of dimension 1 is obvious (one only needs to look at Γ as an abstract graph). Moreover the boundaries of the discs of P (with their correct orientation) are reconstructed by the rules given in Fig. 1.3, so the action of ∂ on 2-cells can also be described in a very explicit way. We avoid introducing specific notations, because it should be clear anyway to the reader how to proceed in the computation in any specific example. We confine ourselves to the following result which has been used in Section 4.6; in its statement we will use the terminology of type of an edge for a region introduced after Definition 4.4.6.

Lemma 10.1.1. $H^2(P;\mathbb{Z})$ is the quotient of the free $\mathbb{Z}$-module generated by a symbol $\hat{\Delta}$ for every region Δ of P, under the relations

$$\hat{\Delta}^{(e)} - \hat{\Delta}^{(e)}_+ + \hat{\Delta}^{(e)}_-,$$

as e varies among the edges of P and $\Delta^{(e)}$, $\Delta^{(e)}_+$, $\Delta^{(e)}_-$ are the regions for which e is respectively of type I, II_+, II_-.

When one considers manifolds with boundary, a standard spine is a deformation retract of the corresponding manifold, so the homology and cohomology groups are isomorphic (and moreover the isomorphism is canonical if one considers an embedded standard spine). We only want to recall, since this is used in Sections 4.6 and 6.1, that when M is an oriented manifold with non-empty boundary there is a canonical isomorphism of $H_1(M, \partial M; \mathbb{Z})$ with $H^2(M; \mathbb{Z})$, obtained by representing elements of $H_1(M, \partial M; \mathbb{Z})$ by means of properly embedded oriented 1-submanifolds, and defining their action on 2-chains as the algebraic intersection number. If M has a branched standard spine P this duality can be explicitly described rather easily. Namely, if $\pi : M \to P$ is the retraction, to an element $[n_1 \hat{\Delta}_1 + \cdots + n_k \hat{\Delta}_k]$ of $H^2(P; \mathbb{Z})$ we associate the class in $H_1(M, \partial M; \mathbb{Z})$ of the properly embedded 1-submanifold obtained by taking $|n_i|$ distinct points $\{x_j^i\}$ inside Δ_i and orienting the segments $\pi^{-1}(x_j^i)$ coherently with the flow or not, depending on the sign of n_i.

Now we turn to the case of a closed manifold M, which we represent by means of a branched standard spine P whose associated manifold has boundary S^2 with trivial bicoloration. We can extend the cellularization of P to one of M by adding one 3-cell only; it is quite easy to see that (regardless of the orientation) ∂ is null on the 3-cell. Therefore also in this case the homology and cohomology can be computed very easily starting from the normal o-graph Γ of P. We will show now how to express directly on P the dualities which exist in this case.

Let us introduce more specific notations for the cellularization of M. Denote by $e_1, \ldots, e_{2n}$ the edges of Γ, with their natural orientation, and by $\Delta_1, \ldots, \Delta_{n+1}$ the discs of P, oriented as prescribed by the branching (here n is the number of vertices of Γ).

Proposition 10.1.2. *The duality isomorphism* $H_1(M; \mathbb{Z}) \to H^2(M; \mathbb{Z})$ *is defined as*

$$\left[\sum_{i=1}^{n} k_i e_i \right] \mapsto \left[\sum_{i=1}^{n} k_i f(e_i) \right],$$

where for an edge e the 2-cochain $f(e)$ is a sum of a contribution for each of the two ends of e, and, with the notation of Fig. 10.1, each end contributes as follows:

$$a_1, a_2, a_3 \mapsto 0, \qquad a_4 \mapsto -\hat{A}_4, \qquad b_1, b_2, b_3 \mapsto 0, \qquad b_4 \mapsto \hat{B}_4.$$

Similarly the inverse isomorphism is induced by the map which to a disc associates a 1-cochain according to:

$$A_1, A_2, A_3, A_5, A_6 \mapsto 0, \qquad A_4 \mapsto -\hat{a}_4, \qquad B_1, B_2, B_3, B_5, B_6 \mapsto 0, \qquad B_4 \mapsto \hat{b}_4.$$

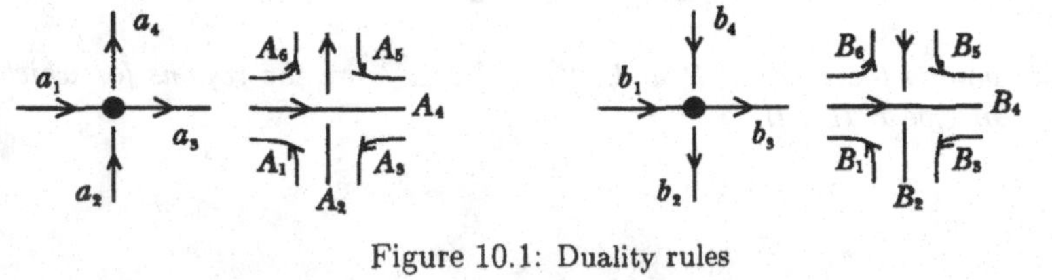

Figure 10.1: Duality rules

Proof of 10.1.2. The idea is just that a cellular 1-chain should be pushed off P towards the positive direction of the flow carried by P. This makes the 1-chain transversal to the 2-cells, and hence we only have to determine the signs of the intersections. We leave to the reader to check that the rules which we have given actually translate this process. $\boxed{10.1.2}$

Of course the isomorphism $H_2(M;\mathbb{Z}) \cong H^1(M;\mathbb{Z})$ is obtained essentially by the same formulae (one should just take the dual of every symbol).

10.2 More homological invariants

In this section we provide effective methods of computation of two homological invariants directly from normal o-graphs. Given a closed oriented manifold M we will show how to compute the Bockstein homomorphism $H^1(M;\mathbb{Z}_2) \to H^2(M;\mathbb{Z})$, and explain that this defines a lifting to $\mathbb{Z}$ of the operation $H^1(M;\mathbb{Z}_2) \to H^2(M;\mathbb{Z}_2)$ of taking the square of an element. When M has a spin structure we will show how to compute a map $H^1(M;\mathbb{Z}_2) \to \mathbb{Z}_4$ (a lifting to $\mathbb{Z}_4$ of the operation $H^1(M;\mathbb{Z}_2) \to H^3(M;\mathbb{Z}_2) = \mathbb{Z}_2$ of taking the cube of an element). It is remarkable that in both cases the invariance under the moves of the calculus could be established in a very elementary way (since in both cases the invariants are classical ones, we will not do this explicitly).

Let us fix a branched standard spine P of M minus a 3-ball, with trivial bicoloration on the boundary, and the normal o-graph Γ which represents P. By duality we identify $H^1(M;\mathbb{Z}_2)$ with $H_2(M;\mathbb{Z}_2)$; since only one puncture has been made in M, one easily sees that the inclusion $P \hookrightarrow M$ induces an isomorphism of $H_2(P;\mathbb{Z}_2)$ onto $H_2(M;\mathbb{Z}_2)$; moreover $H_2(P;\mathbb{Z}_2)$ can be canonically identified with the set of closed surfaces contained in P (because every such a surface is automatically a union of discs of P).

Given a surface Σ contained in P we denote by $s(\Sigma)$ the singular set (in other words, the bending locus) of Σ, namely the union of the edges e such that Σ contains two discs which, with the orientation given by the branching of P, induce on e the same orientation. The following fact is straight-forward:

Lemma 10.2.1. *The natural orientation of the edges of Γ turns $s(\Sigma)$ into a disjoint union of oriented loops.*

This lemma implies that $s(\Sigma)$ defines an element of $H_1(P;\mathbb{Z}) \cong H_1(M;\mathbb{Z}) \cong H^2(M;\mathbb{Z})$.

Proposition 10.2.2. *The map*

$$H_2(M;\mathbb{Z}_2) \ni [\Sigma] \mapsto [s(\Sigma)] \in H_1(M;\mathbb{Z})$$

is the Bockstein homomorphism.

Proof of 10.2.2. Denote by b the Bockstein homomorphism and recall that $b([\Sigma])$ is obtained by taking a $\mathbb{Z}$-lifting $\tilde{\Sigma}$ of Σ and defining $b([\Sigma]) = [\gamma]$ if $\partial\tilde{\Sigma} = 2\gamma$. We take $\tilde{\Sigma}$ to be the chain of discs in P which belong to Σ, all with the orientation given by the branching of P. The conclusion easily follows from the definitions. $\boxed{10.2.2}$

Proposition 10.2.3. *Consider the Bockstein homomorphism $H^1(M; \mathbb{Z}_2) \to H^2(M; \mathbb{Z})$ and compose it with the reduction modulo 2 of coefficients $H^2(M; \mathbb{Z}) \to H^2(M; \mathbb{Z}_2)$. The result is the map*

$$H^1(M; \mathbb{Z}_2) \ni \xi \mapsto \xi^2 \in H^2(M; \mathbb{Z}_2)$$

where $\xi^2 = \xi \cdot \xi$ is the cohomology cup product.

Proof of 10.2.3. This is a very general fact the proof of which becomes particularly clear when carried out using a branched standard spine P as above. If $[\Sigma] \in H_2(P; \mathbb{Z}_2)$ represents ξ by duality then we can obtain a homologous transversal copy $\tilde{\Sigma}$ of Σ in M by pushing Σ towards the positive direction of the flow defined by P. Then ξ^2 is represented by $\Sigma \cap \tilde{\Sigma}$; moreover $\Sigma \cap \tilde{\Sigma}$ is homologous to $s(\Sigma)$, and the conclusion follows. $\boxed{10.2.3}$

Now we can show that using a branched standard spine P as above we can also explicitly compute the map

$$H^1(M; \mathbb{Z}_2) \ni \xi \mapsto \xi^3 \in H^3(M; \mathbb{Z}_2) \cong \mathbb{Z}_2.$$

Later we will show that in presence of a spin structure a lifting to $\mathbb{Z}_4$ of this map is possible.

Proposition 10.2.4. *If $[\Sigma] \in H_2(P; \mathbb{Z}_2)$ is dual to ξ then ξ^3 is the number modulo 2 of vertices at which $s(\Sigma)$, viewed as an oriented curve in Γ, turns left.*

Proof of 10.2.4. One sees that each connected component of $s(\Sigma)$ contributes with 1 to ξ^3 if a regular neighbourhood N of the component in Σ is a Möbius strip, and it contributes with 0 if N is a cylinder. The normal o-graph Γ allows to see N in 3-space as an almost always horizontal ribbon; moreover there are some half-twists which exactly correspond to left-turns of $s(\Sigma)$. $\boxed{10.2.4}$

Let us now consider the case where M has a certain spin structure σ, defined by some $\mathbb{Z}_2$-colouring of the edges of a normal o-graph Γ as above; we will denote by $\tilde{\sigma}$ this $\mathbb{Z}_2$-colouring, and view $\tilde{\sigma}$ as a 1-cochain with coefficients in $\mathbb{Z}_2$. In such a case, given a closed surface Σ in P, one can consider a regular neighbourhood N of $s(\Sigma)$ in Σ, and compute the number of half-twists of the ribbon N with respect to σ. This number is only defined modulo 4, and we will denote it by $k(\xi) \in \mathbb{Z}_4$, where $\xi \in H^1(M; \mathbb{Z}_2)$ is dual to $[\Sigma]$. The reduction modulo 2 of $k(\xi)$ is ξ^3 (see also the proof of the previous proposition).

Here is an abstract version of the definition of $k(\xi)$: we consider a (smooth) surface Σ embedded in M, representing a class in $H_2(M; \mathbb{Z}_2)$; we consider another copy of Σ transversal to the original one, so that the intersection is a (smooth) curve embedded in Σ; this embedding defines a framing on the curve, and we can compute the number (well-defined modulo 4) of half-twists of this framing with respect to the spin structure. The proof that the number thus defined really depends only on $[\Sigma] \in H_2(M; \mathbb{Z}_2)$ can be obtained using for instance the techniques of [10], [32].

Proposition 10.2.5. *If $\xi \in H^1(M; \mathbb{Z}_2)$ is dual to $[\Sigma]$ then $k(\xi)$ is given by a sum $\eta(\Sigma) + \vartheta(\Sigma) + \tau(\Sigma)$, where:*

- $\eta(\Sigma)$ *is the reduction modulo 4 of the algebraic sum of the indices of vertices of* Γ *at which* $s(\Sigma)$ *turns left.*

- $\vartheta(\Sigma)$ *is twice the value of the cochain* $\bar{\sigma}$ *on the chain* $s(\Sigma)$ *(here 'twice' means that we embed* $\mathbb{Z}_2$ *into* $\mathbb{Z}_4$ *via* $0 \mapsto 0$, $1 \mapsto 2$*);*

- $\tau(\Sigma)$ *is the reduction modulo 4 of twice the number of components of* $s(\Sigma)$.

Proof of 10.2.5. Let N be the ribbon in Σ with core $s(\Sigma)$. Consider a component γ_0 of $s(\Sigma)$, the corresponding ribbon $N_0 \subset N$ and a ribbon $\tilde{N}_0$ which has the same core γ_0 as N_0, is homeomorphic to a cylinder and defines a framing along γ_0 which does not lift to the principal Spin(3)-bundle which defines σ. Then the contribution of N_0 to $k(\xi)$ is the number (well defined modulo 4) of half-twists of N_0 with respect to $\tilde{N}_0$.

If γ_0 never turns left and does not contain any edge with colour 1, then its contribution is 2, because N_0 is a cylinder and the framing it defines lifts to the principal Spin(3)-bundle; this explains the τ-part of k. It is now quite obvious that each edge with colour 1 in γ_0 will change the contribution by 2; this justifies the ϑ-part of k. To conclude one has to check that the rest of k is the η-part; namely one has to show that the contribution to k of every left-turn of γ_0 is given by the index of the vertex at which the left-turn takes place. This can be checked by very simple figures, and we leave it to the reader. $\boxed{10.2.5}$

10.3 Evenly framed knots in a spin manifold

Let us consider a fixed closed spin 3-manifold (M, σ), a branched standard spine P of the punctured version of M with trivial bicoloration on the boundary S^2, and the normal o-graph Γ which represents P; we will use the fact that Γ, as an abstract graph, is identified to the singular set of P. We also consider a $\mathbb{Z}_2$-colouring of the edges of Γ which encodes the spin structure σ, according to Theorem 1.4.4.

We want to determine which framed knots are even in (M, σ). Here "even" means that the framing does *not* lift to the principal Spin(3)-bundle which defines σ. To this end we consider a regular neighbourhood U of Γ in P, and we note that any knot in M can a isotoped to a C^1 curve in U (this is because the smooth components of P are discs). Moreover a C^1 knot γ in U is naturally framed by the triple (γ', n, v), where n is the positive normal to γ' in P and v is the vector field carried by P. So we only have to determine whether the natural framing of a C^1 knot in U is even or odd.

Now we note that given a C^1 knot γ in U, up to removing some curls, we can assume that γ is always roughly parallel to the singular set $S(\Gamma)$, except at vertices where γ has to turn. We might also assume that γ never follows an edge and immediately afterwards the same edge in the opposite direction. Therefore γ determines a locally injective simplicial loop $\tilde{\gamma}$ in Γ. (It is not quite true that $\tilde{\gamma}$ determines γ, because self-crossings of $\tilde{\gamma}$ can be loosened in two different ways; however two C^1 knots γ_1, γ_2 in U such that $\tilde{\gamma}_1 = \tilde{\gamma}_2$ will have both even or both odd natural framing with respect to σ).

So, let us consider a loop $\tilde{\gamma} : [0, 1] \to \Gamma$ which is locally injective and simplicial, with $\tilde{\gamma}(t)$ vertex if and only if t is one of the values $0 = t_0 < t_1 < \cdots < t_k = 1$. Let $\varepsilon_i \in \{0, 1\}$ be the colour of the edge $\gamma([t_{i-1}, t_i])$, and $\delta_i \in \{0, 1/2\}$ be given by 0 if the edges $\tilde{\gamma}([t_{i-1}, t_i])$ and $\tilde{\gamma}([t_i, t_{i+1}])$ have coherent orientation at $\tilde{\gamma}(t_i)$, 1/2 otherwise.

Here indices are understood modulo k. We define $\phi(\tilde{\gamma}) = \sum_{i=1}^{k}(\varepsilon_i + \delta_i)$. Let γ be a C^1 knot in U parallel to $S(\Gamma)$ and such that the corresponding simplicial loop is $\tilde{\gamma}$.

Proposition 10.3.1. $\phi(\tilde{\gamma})$ *is an integer and the natural framing of γ is even if and only if $\phi(\tilde{\gamma})$ is odd.*

Proof of 10.3.1. We only give a sketch without figures, leaving the details to the reader. Consider γ as a slight modification of $\tilde{\gamma}$ at vertices. Next take the tangent field ν to P associated to the $\mathbb{Z}_2$-colouring (recall that ν is almost everywhere transverse to the edges of Γ and points towards the 1-sheeted side of the edge, except that there is a curl along every edge with colour 1). By definition the framing is even if and only if the degree modulo 2 of γ' with respect to ν is 1. So we consider the points of Γ where γ' and ν are parallel. It is easy to see that if $\varepsilon_i = 1$ then on $\gamma([t_{i-1}, t_i])$ there is exactly one point where $\gamma' = \nu$ and one where $\gamma' = -\nu$; if $\varepsilon_i = 0$ then on $\gamma([t_{i-1}, t_i])$ there is no point where γ' is parallel to ν. Similarly near $\gamma(t_i)$ there is a point where $\gamma' = \pm\nu$ if and only if $\delta_i = 1/2$, and if so there is exactly one. Moreover all the points where $\gamma' = \pm\nu$ are regular, and the conclusion follows quite easily. $\boxed{10.3.1}$

Bibliography

[1] M. ATIYAH, *On framings of 3-manifolds*, Topology **29** (1990), 1-7.

[2] D. BENNEQUIN, *Entrelacements et équations de Pfaff*, Astérisque **107-108** (1983).

[3] C. BLANCHET, *Invariants of three-manifolds with spin structure*, Comment. Math. Helvetici **67** (1992), 406-427.

[4] C. BLANCHET, N. HABEGGER, G. MASBAUM, P. VOGEL, *Three-manifold invariants derived from the Kauffman bracket*, Topology **31** (1992), 685-699.

[5] R. BENEDETTI, *A combinatorial approach to combings and framings on 3-manifolds*, to appear in the Proceedings of the Second European Congress of Mathematics, Budapest, 1996.

[6] R. BENEDETTI – C. PETRONIO, *A Finite graphic calculus for 3-manifolds*, Manuscripta Math. **88** (1995), 291-310.

[7] R. BENEDETTI – C. PETRONIO, *On Roberts' proof of the Turaev-Walker theorem*, J. Knot Theory Ramif. **5** (1996), 427-439.

[8] R. BENEDETTI – C. PETRONIO, Lectures on Hyperbolic Geometry, Springer-Verlag, Berlin-Heidelberg-New York, 1992.

[9] R. BENEDETTI – J.-J. RISLER, Real algebraic and semi-algebraic sets, Actualités Mathématiques, Hermann, Paris, 1990.

[10] R. BENEDETTI – R. SILHOL, *Spin and Pin⁻ structures, immersed and embedded surfaces and a result of Segre on real cubic surfaces*, Topology **34** (1995), 651-678.

[11] J. CHRISTY, *Branched surfaces and attractors I*, Trans. Amer. Math. Soc. **336** (1993), 759-784.

[12] J. CHRISTY, *Standard spines and branched surfaces*, in preparation.

[13] B. G. CASLER, *An imbedding theorem for connected 3-manifolds with boundary*. Proc. Amer. Math. Soc. **16** (1965), 559-566.

[14] YA. ELIASHBERG, *Contact 3-manifolds twenty years after J. Martinet's work*, Ann. Inst. Fourier (Grenoble) **42** (1992), 165-192.

[15] YA. ELIASHBERG, *Classification of overtwisted contact structures*, Invent. Math. **98** (1989), 623-637.

[16] YA. ELIASHBERG, W. P. THURSTON, *Confoliations*, Preprint (1996).

[17] W. FLOYD – U. OERTEL, *Incompressible surfaces via branched surfaces*, Topology **23** (1984), 117-125.

[18] D. GABAI, *Foliations and the types of 3-manifolds*, J. Differential Geometry **18** (1983), 445-503.

[19] D. GABAI – U. OERTEL, *Essential laminations in 3-manifolds*, Ann. of Math. **130** (1989), 41-73.

[20] D. GILLMAN – D. ROLFSEN, *The Zeeman conjecture for standard spines is equivalent to the Poincaré conjecture*, Topology **22** (1983), 315-323.

[21] D. GILLMAN – D. ROLFSEN, *Three-manifolds embed in small 3-complexes*, Int. J. Math. **3** (1992), 179-183.

[22] E. GIROUX, *Convexité en topologie de contact*, Comm. Math. Helv. **66** (1991), 637-677.

[23] M. A. HENNINGS, *Invariants of links and 3-manifolds obtained from Hopf algebras*, Preprint (1989).

[24] I. ISHII, *Flows and spines*, Tokyo J. Math. **9** (1986), 505-525.

[25] I. ISHII, *Combinatorial construction of a non-singular flow on a 3-manifold*, Kobe J. Math. **3** (1986), 201-208.

[26] I. ISHII, *Moves for flow-spines and topological invariants of 3-manifolds*, Tokyo J. Math. **15** (1992), 297-312.

[27] S. J. KAPLAN, *Constructing framed 4-manifolds with given almost framed boundaries*, Trans. Amer. Math. Soc. **254** (1979), 237-263.

[28] L. H. KAUFFMAN - S. LINS, Temperley-Lieb recoupling theory and invariants of 3-manifolds, Princeton University Press, Princeton NJ, 1994.

[29] L. H. KAUFFMAN - D. E. RADFORD *Invariants of 3-manifolds derived from finite dimensional Hopf algebras*, J. Knot. Theory Ramif. **4** (1995), 131-162.

[30] R. C. KIRBY , The Topology of 4-manifolds, Lectures Notes in Mathematics 1374, Springer-Verlag, Berlin-Heidelberg-New York, 1989.

[31] R. C. KIRBY, MELVIN, *The 3-manifolds invariants of Witten and Reshetikhin-Turaev for sl(2,$\mathbb{C}$)*, Invent. Math. **105** (1991), 473-545.

[32] R. C. KIRBY – L. R. TAYLOR, *Pin structures on low-dimensional manifolds*, In: "Geometry of low-dimensional manifolds", Vol. 2, S. K. Donaldson and C. D. Thomas Eds., Cambridge University Press, Cambridge, 1989, pp. 177-241.

[33] G. KUPENBERG, *Non-involutory Hopf-algebras and 3-manifold invariants*, Preprint (1995).

[34] W. B. R. Lickorish, *Three-manifolds and the Temperley-Lieb algebra*, Math. Ann. **290** (1991), 657-670.

[35] R. LUTZ, *Structures de contact sur les fibrés principaux en cercles de dimension 3*, Ann. Inst. Fourier (Grenoble) **27** (1977), 1-15.

[36] J. MARTINET, *Formes de contact sur les varietès de dimension 3*, In: Lecture Notes in Math. 209, Springer-Verlag, Berlin-Heidelberg-New York, 1971, pp. 142-163.

[37] S. V. MATVEEV, *Transformations of special spines and the Zeeman conjecture*, Math. USSR-Izv. **31** (1988), 423-434.

[38] S. V. MATVEEV – A. T. FOMENKO, *Constant energy surfaces of Hamiltonian systems, enumeration of three-dimensional manifolds in increasing order of complexity, and computation of volumes of closed hyperbolic manifolds*, Russ. Math. Surv. **43** (1988), 3-25.

[39] J. W. MILNOR, Topology from the differentiable viewpoint. The University Press of Virginia, Charlottesville VA, 1965.

[40] J. W. MILNOR – J. D. STASHEFFS, Characteristic classes. Princeton University Press, Princeton NJ, 1974.

[41] S. P. NOVIKOV, *Topology of foliations*, Trans. Moscow Math. Soc. **14** (1965), 268-304.

[42] U. OERTEL, *Measured laminations in 3-manifolds*, Trans. Amer. Math. Soc. **305** (1988), 531-573.

[43] C. PETRONIO, *Standard spines and 3-manifolds*, Tesi di Perfezionamento, Scuola Normale Superiore, Pisa, 1995.

[44] C. PETRONIO, *Ideal triangulations of link complements and hyperbolicity equations*. To appear in Geom. Ded.

[45] R. PIERGALLINI, *Standard moves for standard polyhedra and spines*. Rendiconti Circ. Mat. Palermo **37**, suppl. 18 (1988), 391-414.

[46] N. YA. RESHETIKHIN – V. G. TURAEV, *Invariants of three-manifolds via link polynomials and quantum groups*, Invent. Math. **103** (1991), 547-597.

[47] J. D. ROBERTS *Skein theory and Turaev-Viro invariants*, Topology **34** (1995), 771-787.

[48] J. D. ROBERTS *Refined state sum invariants of 3- and 4-manifolds*, Proc. 1993 Georgia Topology Conference.

[49] H. ROSENBERG, *Foliations by planes*, Topology **6** (1967), 131-138.

[50] C. P. ROURKE, *A new proof that Ω_3 is zero*, L. London Math. Soc. **31** (1985), 373-376.

[51] C. P. ROURKE – B. J. SANDERSON, An introduction to piecewise linear topology. Ergebn. der Math. Bd 69, Springer-Verlag, Berlin-Heidelberg-New York, 1982.

[52] E. H. SPANIER Algebraic topology, McGraw-Hill, New York, 1966.

[53] W. P. THURSTON, *Norm on the homology of 3-manifolds*, Mem. Amer. Math. Soc. **339** (1986), 99-130.

[54] D. TIESCHLER, *On fibering certain foliated manifolds over S^1*, Topology **9** (1970), 153-154.

[55] V. G. TURAEV, *Euler structures, nonsingular vector fields, and torsion of Reidemeister type*, Math. USSR-Izv. **34** (1990), 627-662.

[56] V. G. TURAEV, Quantum Invariants of Knots and 3-manifolds, Studies in Math. 18, de Gruyter, Berlin, 1994.

[57] V. G. TURAEV – O. YA. VIRO, *State sum invariants of 3-manifolds and quantum 6j-symbols*, Topology **31** (1992), 865-902.

[58] J. R. WEEKS, *SnapPea*, the hyperbolic structures computer program, latest release, 1996.

[59] J. WOOD, *Foliations on 3-manifolds*, Ann. Math. **89** (1969), 336-358.

[60] R. F. WILLIAMS, *Expanding attractors*, Publ. Math. I.H.E.S. **43** (1974), 169-203.

[61] E. WITTEN, *Quantum field theory and the Jones polynomial*, Comm. Math. Phys. **121** (1989), 351-399.

Index

balanced Hopf algebra, 113
bicoloration of the boundary, 28
black region, 42
blow-up, 34
branched standard spine, 3
branching, 23
    oriented —, 25
bumping move, 61

c_P, 86
combinatorial
    — Pontrjagin move, 81
    — realization, 1
combing, 2
comparison class, 74
complexity, 56
concave
    — flow, 42
    — point of a flow, 40
contour curve, 44
convex
    — flow, 45
    — point of a flow, 40

desingularization, 34

embedded
    — move, 47
    — o-graph, 47
    — spine, 44
Euler cochain, 85

$\mathcal{F}_i(P)$, 26
flow
    smooth generic —, 40
    topological generic —, 42
flow-spine, 64
framing, 2
fundamental chain, 26

$\mathcal{G}$, 6
$\mathcal{G}_{\text{comb}}$, 6
$\mathcal{G}_{\text{fram}}$, 7
$\mathcal{G}_{\text{spin}}$, 7

homotopy through concave
    traversing flows, 47
Hopf
    — algebra, 111
    — number, 80

index of a vertex, 27

$L(\Gamma)$, 20

$\widehat{M}$, 20
Matveev-Piergallini move, 14
    — on branched spines, 37
$\mathcal{M}$, 2
$\mathcal{M}_{\text{comb}}$, 2
$\mathcal{M}_{\text{fram}}$, 2
$\mathcal{M}_{\text{spin}}$, 2

$\mathcal{N}$, 6
normal
    — section, 65
    — triple, 64
normal o-graph, 6, 26
    closed —, 6
    framed —, 7
    spin —, 7

o-graph, 15
orbit of a flow, 42
oriented branching, 25
    — on a simple polyhedron, 51
oriented simple polyhedron, 50
oriented standard polyhedron, 14

Pontrjagin move, 76
positive move, 34

reconstruction map, 1
recoupling theory, 99
Roberts link, 20
RTW^r, $RTW^{r(0)}$, $RTW^{r(2)}$, 100

separating curve, 42
simple polyhedron, 13

sliding moves
    pure —, 51
    simple —, 51
    standard —, 47
spin structure, 2
spine, 14
standard polyhedron, 13

transversal, flow — to itself, 42
traversing flow, 42
TV^r, $TV^{r(0)}$, $TV^{r(2)}$, 101
$TV_0^{r(0)}$, $TV_1^{r(0)}$, $TV_2^{r(0)}$, 104
type of an edge for a region, 52

weight system, 116
white region, 42

Lecture Notes in Mathematics

For information about Vols. 1–1469
please contact your bookseller or Springer-Verlag

Vol. 1470: E. Odell, H. Rosenthal (Eds.), Functional Analysis. Proceedings, 1987-89. VII, 199 pages. 1991.

Vol. 1471: A. A. Panchishkin, Non-Archimedean L-Functions of Siegel and Hilbert Modular Forms. VII, 157 pages. 1991.

Vol. 1472: T. T. Nielsen, Bose Algebras: The Complex and Real Wave Representations. V, 132 pages. 1991.

Vol. 1473: Y. Hino, S. Murakami, T. Naito, Functional Differential Equations with Infinite Delay. X, 317 pages. 1991.

Vol. 1474: S. Jackowski, B. Oliver, K. Pawałowski (Eds.), Algebraic Topology, Poznań 1989. Proceedings. VIII, 397 pages. 1991.

Vol. 1475: S. Busenberg, M. Martelli (Eds.), Delay Differential Equations and Dynamical Systems. Proceedings, 1990. VIII, 249 pages. 1991.

Vol. 1476: M. Bekkali, Topics in Set Theory. VII, 120 pages. 1991.

Vol. 1477: R. Jajte, Strong Limit Theorems in Noncommutative L_2-Spaces. X, 113 pages. 1991.

Vol. 1478: M.-P. Malliavin (Ed.), Topics in Invariant Theory. Seminar 1989-1990. VI, 272 pages. 1991.

Vol. 1479: S. Bloch, I. Dolgachev, W. Fulton (Eds.), Algebraic Geometry. Proceedings, 1989. VII, 300 pages. 1991.

Vol. 1480: F. Dumortier, R. Roussarie, J. Sotomayor, H. Żołądek, Bifurcations of Planar Vector Fields: Nilpotent Singularities and Abelian Integrals. VIII, 226 pages. 1991.

Vol. 1481: D. Ferus, U. Pinkall, U. Simon, B. Wegner (Eds.), Global Differential Geometry and Global Analysis. Proceedings, 1991. VIII, 283 pages. 1991.

Vol. 1482: J. Chabrowski, The Dirichlet Problem with L^2-Boundary Data for Elliptic Linear Equations. VI, 173 pages. 1991.

Vol. 1483: E. Reithmeier, Periodic Solutions of Nonlinear Dynamical Systems. VI, 171 pages. 1991.

Vol. 1484: H. Delfs, Homology of Locally Semialgebraic Spaces. IX, 136 pages. 1991.

Vol. 1485: J. Azéma, P. A. Meyer, M. Yor (Eds.), Séminaire de Probabilités XXV. VIII, 440 pages. 1991.

Vol. 1486: L. Arnold, H. Crauel, J.-P. Eckmann (Eds.), Lyapunov Exponents. Proceedings, 1990. VIII, 365 pages. 1991.

Vol. 1487: E. Freitag, Singular Modular Forms and Theta Relations. VI, 172 pages. 1991.

Vol. 1488: A. Carboni, M. C. Pedicchio, G. Rosolini (Eds.), Category Theory. Proceedings, 1990. VII, 494 pages. 1991.

Vol. 1489: A. Mielke, Hamiltonian and Lagrangian Flows on Center Manifolds. X, 140 pages. 1991.

Vol. 1490: K. Metsch, Linear Spaces with Few Lines. XIII, 196 pages. 1991.

Vol. 1491: E. Lluis-Puebla, J.-L. Loday, H. Gillet, C. Soulé, V. Snaith, Higher Algebraic K-Theory: an overview. IX, 164 pages. 1992.

Vol. 1492: K. R. Wicks, Fractals and Hyperspaces. VIII, 168 pages. 1991.

Vol. 1493: E. Benoît (Ed.), Dynamic Bifurcations. Proceedings, Luminy 1990. VII, 219 pages. 1991.

Vol. 1494: M.-T. Cheng, X.-W. Zhou, D.-G. Deng (Eds.), Harmonic Analysis. Proceedings, 1988. IX, 226 pages. 1991.

Vol. 1495: J. M. Bony, G. Grubb, L. Hörmander, H. Komatsu, J. Sjöstrand, Microlocal Analysis and Applications. Montecatini Terme, 1989. Editors: L. Cattabriga, L. Rodino. VII, 349 pages. 1991.

Vol. 1496: C. Foias, B. Francis, J. W. Helton, H. Kwakernaak, J. B. Pearson, H_∞-Control Theory. Como, 1990. Editors: E. Mosca, L. Pandolfi. VII, 336 pages. 1991.

Vol. 1497: G. T. Herman, A. K. Louis, F. Natterer (Eds.), Mathematical Methods in Tomography. Proceedings 1990. X, 268 pages. 1991.

Vol. 1498: R. Lang, Spectral Theory of Random Schrödinger Operators. X, 125 pages. 1991.

Vol. 1499: K. Taira, Boundary Value Problems and Markov Processes. IX, 132 pages. 1991.

Vol. 1500: J.-P. Serre, Lie Algebras and Lie Groups. VII, 168 pages. 1992.

Vol. 1501: A. De Masi, E. Presutti, Mathematical Methods for Hydrodynamic Limits. IX, 196 pages. 1991.

Vol. 1502: C. Simpson, Asymptotic Behavior of Monodromy. V, 139 pages. 1991.

Vol. 1503: S. Shokranian, The Selberg-Arthur Trace Formula (Lectures by J. Arthur). VII, 97 pages. 1991.

Vol. 1504: J. Cheeger, M. Gromov, C. Okonek, P. Pansu, Geometric Topology: Recent Developments. Editors: P. de Bartolomeis, F. Tricerri. VII, 197 pages. 1991.

Vol. 1505: K. Kajitani, T. Nishitani, The Hyperbolic Cauchy Problem. VII, 168 pages. 1991.

Vol. 1506: A. Buium, Differential Algebraic Groups of Finite Dimension. XV, 145 pages. 1992.

Vol. 1507: K. Hulek, T. Peternell, M. Schneider, F.-O. Schreyer (Eds.), Complex Algebraic Varieties. Proceedings, 1990. VII, 179 pages. 1992.

Vol. 1508: M. Vuorinen (Ed.), Quasiconformal Space Mappings. A Collection of Surveys 1960-1990. IX, 148 pages. 1992.

Vol. 1509: J. Aguadé, M. Castellet, F. R. Cohen (Eds.), Algebraic Topology - Homotopy and Group Cohomology. Proceedings, 1990. X, 330 pages. 1992.

Vol. 1510: P. P. Kulish (Ed.), Quantum Groups. Proceedings, 1990. XII, 398 pages. 1992.

Vol. 1511: B. S. Yadav, D. Singh (Eds.), Functional Analysis and Operator Theory. Proceedings. 1990. VIII, 223 pages. 1992.

Vol. 1512: L. M. Adleman, M.-D. A. Huang, Primality Testing and Abelian Varieties Over Finite Fields. VII, 142 pages. 1992.

Vol. 1513: L. S. Block, W. A. Coppel, Dynamics in One Dimension. VIII, 249 pages. 1992.

Vol. 1514: U. Krengel, K. Richter, V. Warstat (Eds.), Ergodic Theory and Related Topics III. Proceedings, 1990. VIII, 236 pages. 1992.

Vol. 1515: E. Ballico, F. Catanese, C. Ciliberto (Eds.), Classification of Irregular Varieties. Proceedings, 1990. VII, 149 pages. 1992.

Vol. 1516: R. A. Lorentz, Multivariate Birkhoff Interpolation. IX, 192 pages. 1992.

Vol. 1517: K. Keimel, W. Roth, Ordered Cones and Approximation. VI, 134 pages. 1992.

Vol. 1518: H. Stichtenoth, M. A. Tsfasman (Eds.), Coding Theory and Algebraic Geometry. Proceedings, 1991. VIII, 223 pages. 1992.

Vol. 1519: M. W. Short, The Primitive Soluble Permutation Groups of Degree less than 256. IX, 145 pages. 1992.

Vol. 1520: Yu. G. Borisovich, Yu. E. Gliklikh (Eds.), Global Analysis – Studies and Applications V. VII, 284 pages. 1992.

Vol. 1521: S. Busenberg, B. Forte, H. K. Kuiken, Mathematical Modelling of Industrial Process. Bari, 1990. Editors: V. Capasso, A. Fasano. VII, 162 pages. 1992.

Vol. 1522: J.-M. Delort, F. B. I. Transformation. VII, 101 pages. 1992.

Vol. 1523: W. Xue, Rings with Morita Duality. X, 168 pages. 1992.

Vol. 1524: M. Coste, L. Mahé, M.-F. Roy (Eds.), Real Algebraic Geometry. Proceedings, 1991. VIII, 418 pages. 1992.

Vol. 1525: C. Casacuberta, M. Castellet (Eds.), Mathematical Research Today and Tomorrow. VII, 112 pages. 1992.

Vol. 1526: J. Azéma, P. A. Meyer, M. Yor (Eds.), Séminaire de Probabilités XXVI. X, 633 pages. 1992.

Vol. 1527: M. I. Freidlin, J.-F. Le Gall, Ecole d'Eté de Probabilités de Saint-Flour XX – 1990. Editor: P. L. Hennequin. VIII, 244 pages. 1992.

Vol. 1528: G. Isac, Complementarity Problems. VI, 297 pages. 1992.

Vol. 1529: J. van Neerven, The Adjoint of a Semigroup of Linear Operators. X, 195 pages. 1992.

Vol. 1530: J. G. Heywood, K. Masuda, R. Rautmann, S. A. Solonnikov (Eds.), The Navier-Stokes Equations II – Theory and Numerical Methods. IX, 322 pages. 1992.

Vol. 1531: M. Stoer, Design of Survivable Networks. IV, 206 pages. 1992.

Vol. 1532: J. F. Colombeau, Multiplication of Distributions. X, 184 pages. 1992.

Vol. 1533: P. Jipsen, H. Rose, Varieties of Lattices. X, 162 pages. 1992.

Vol. 1534: C. Greither, Cyclic Galois Extensions of Commutative Rings. X, 145 pages. 1992.

Vol. 1535: A. B. Evans, Orthomorphism Graphs of Groups. VIII, 114 pages. 1992.

Vol. 1536: M. K. Kwong, A. Zettl, Norm Inequalities for Derivatives and Differences. VII, 150 pages. 1992.

Vol. 1537: P. Fitzpatrick, M. Martelli, J. Mawhin, R. Nussbaum, Topological Methods for Ordinary Differential Equations. Montecatini Terme, 1991. Editors: M. Furi, P. Zecca. VII, 218 pages. 1993.

Vol. 1538: P.-A. Meyer, Quantum Probability for Probabilists. X. 287 pages. 1993.

Vol. 1539: M. Coornaert, A. Papadopoulos, Symbolic Dynamics and Hyperbolic Groups. VIII, 138 pages. 1993.

Vol. 1540: H. Komatsu (Ed.), Functional Analysis and Related Topics, 1991. Proceedings. XXI, 413 pages. 1993.

Vol. 1541: D. A. Dawson, B. Maisonneuve, J. Spencer, Ecole d´Eté de Probabilités de Saint-Flour XXI - 1991. Editor: P. L. Hennequin. VIII, 356 pages. 1993.

Vol. 1542: J.Fröhlich, Th.Kerler, Quantum Groups, Quantum Categories and Quantum Field Theory. VII, 431 pages. 1993.

Vol. 1543: A. L. Dontchev, T. Zolezzi, Well-Posed Optimization Problems. XII, 421 pages. 1993.

Vol. 1544: M.Schürmann, White Noise on Bialgebras. VII, 146 pages. 1993.

Vol. 1545: J. Morgan, K. O'Grady, Differential Topology of Complex Surfaces. VIII, 224 pages. 1993.

Vol. 1546: V. V. Kalashnikov, V. M. Zolotarev (Eds.), Stability Problems for Stochastic Models. Proceedings, 1991. VIII, 229 pages. 1993.

Vol. 1547: P. Harmand, D. Werner, W. Werner, M-ideals in Banach Spaces and Banach Algebras. VIII, 387 pages. 1993.

Vol. 1548: T. Urabe, Dynkin Graphs and Quadrilateral Singularities. VI, 233 pages. 1993.

Vol. 1549: G. Vainikko, Multidimensional Weakly Singular Integral Equations. XI, 159 pages. 1993.

Vol. 1550: A. A. Gonchar, E. B. Saff (Eds.), Methods of Approximation Theory in Complex Analysis and Mathematical Physics IV, 222 pages, 1993.

Vol. 1551: L. Arkeryd, P. L. Lions, P.A. Markowich, S.R. S. Varadhan. Nonequilibrium Problems in Many-Particle Systems. Montecatini, 1992. Editors: C. Cercignani, M. Pulvirenti. VII, 158 pages 1993.

Vol. 1552: J. Hilgert, K.-H. Neeb, Lie Semigroups and their Applications. XII, 315 pages. 1993.

Vol. 1553: J.-L- Colliot-Thélène, J. Kato, P. Vojta. Arithmetic Algebraic Geometry. Trento, 1991. Editor: E. Ballico. VII, 223 pages. 1993.

Vol. 1554: A. K. Lenstra, H. W. Lenstra, Jr. (Eds.), The Development of the Number Field Sieve. VIII, 131 pages. 1993.

Vol. 1555: O. Liess, Conical Refraction and Higher Microlocalization. X, 389 pages. 1993.

Vol. 1556: S. B. Kuksin, Nearly Integrable Infinite-Dimensional Hamiltonian Systems. XXVII, 101 pages. 1993.

Vol. 1557: J. Azéma, P. A. Meyer, M. Yor (Eds.), Séminaire de Probabilités XXVII. VI, 327 pages. 1993.

Vol. 1558: T. J. Bridges, J. E. Furter, Singularity Theory and Equivariant Symplectic Maps. VI, 226 pages. 1993.

Vol. 1559: V. G. Sprindžuk, Classical Diophantine Equations. XII, 228 pages. 1993.

Vol. 1560: T. Bartsch, Topological Methods for Variational Problems with Symmetries. X, 152 pages. 1993.

Vol. 1561: I. S. Molchanov, Limit Theorems for Unions of Random Closed Sets. X, 157 pages. 1993.

Vol. 1562: G. Harder, Eisensteinkohomologie und die Konstruktion gemischter Motive. XX, 184 pages. 1993.

Vol. 1563: E. Fabes, M. Fukushima, L. Gross, C. Kenig, M. Röckner, D. W. Stroock, Dirichlet Forms. Varenna, 1992. Editors: G. Dell'Antonio, U. Mosco. VII, 245 pages. 1993.

Vol. 1564: J. Jorgenson, S. Lang, Basic Analysis of Regularized Series and Products. IX, 122 pages. 1993.

Vol. 1565: L. Boutet de Monvel, C. De Concini, C. Procesi, P. Schapira, M. Vergne. D-modules, Representation Theory, and Quantum Groups. Venezia, 1992. Editors: G. Zampieri, A. D'Agnolo. VII, 217 pages. 1993.

Vol. 1566: B. Edixhoven, J.-H. Evertse (Eds.), Diophantine Approximation and Abelian Varieties. XIII, 127 pages. 1993.

Vol. 1567: R. L. Dobrushin, S. Kusuoka, Statistical Mechanics and Fractals. VII, 98 pages. 1993.

Vol. 1568: F. Weisz, Martingale Hardy Spaces and their Application in Fourier Analysis. VIII, 217 pages. 1994.

Vol. 1569: V. Totik, Weighted Approximation with Varying Weight. VI, 117 pages. 1994.

Vol. 1570: R. deLaubenfels, Existence Families, Functional Calculi and Evolution Equations. XV, 234 pages. 1994.

Vol. 1571: S. Yu. Pilyugin, The Space of Dynamical Systems with the C⁰-Topology. X, 188 pages. 1994.

Vol. 1572: L. Göttsche, Hilbert Schemes of Zero-Dimensional Subschemes of Smooth Varieties. IX, 196 pages. 1994.

Vol. 1573: V. P. Havin, N. K. Nikolski (Eds.), Linear and Complex Analysis – Problem Book 3 – Part I. XXII, 489 pages. 1994.

Vol. 1574: V. P. Havin, N. K. Nikolski (Eds.), Linear and Complex Analysis – Problem Book 3 – Part II. XXII, 507 pages. 1994.

Vol. 1575: M. Mitrea, Clifford Wavelets, Singular Integrals, and Hardy Spaces. XI, 116 pages. 1994.

Vol. 1576: K. Kitahara, Spaces of Approximating Functions with Haar-Like Conditions. X, 110 pages. 1994.

Vol. 1577: N. Obata, White Noise Calculus and Fock Space. X, 183 pages. 1994.

Vol. 1578: J. Bernstein, V. Lunts, Equivariant Sheaves and Functors. V, 139 pages. 1994.

Vol. 1579: N. Kazamaki, Continuous Exponential Martingales and BMO. VII, 91 pages. 1994.

Vol. 1580: M. Milman, Extrapolation and Optimal Decompositions with Applications to Analysis. XI, 161 pages. 1994.

Vol. 1581: D. Bakry, R. D. Gill, S. A. Molchanov, Lectures on Probability Theory. Editor: P. Bernard. VIII, 420 pages. 1994.

Vol. 1582: W. Balser, From Divergent Power Series to Analytic Functions. X, 108 pages. 1994.

Vol. 1583: J. Azéma, P. A. Meyer, M. Yor (Eds.), Séminaire de Probabilités XXVIII. VI, 334 pages. 1994.

Vol. 1584: M. Brokate, N. Kenmochi, I. Müller, J. F. Rodriguez, C. Verdi. Phase Transitions and Hysteresis. Montecatini Terme, 1993. Editor: A. Visintin. VII. 291 pages. 1994.

Vol. 1585: G. Frey (Ed.), On Artin's Conjecture for Odd 2-dimensional Representations. VIII, 148 pages. 1994.

Vol. 1586: R. Nillsen, Difference Spaces and Invariant Linear Forms. XII, 186 pages. 1994.

Vol. 1587: N. Xi, Representations of Affine Hecke Algebras. VIII, 137 pages. 1994.

Vol. 1588: C. Scheiderer, Real and Étale Cohomology. XXIV, 273 pages. 1994.

Vol. 1589: J. Bellissard, M. Degli Esposti, G. Forni, S. Graffi, S. Isola, J. N. Mather, Transition to Chaos in Classical and Quantum Mechanics. Montecatini Terme, 1991. Editor: S. Graffi. VII, 192 pages. 1994.

Vol. 1590: P. M. Soardi, Potential Theory on Infinite Networks. VIII, 187 pages. 1994.

Vol. 1591: M. Abate, G. Patrizio, Finsler Metrics – A Global Approach. IX, 180 pages. 1994.

Vol. 1592: K. W. Breitung, Asymptotic Approximations for Probability Integrals. IX, 146 pages. 1994.

Vol. 1593: J. Jorgenson & S. Lang, D. Goldfeld, Explicit Formulas for Regularized Products and Series. VIII, 154 pages. 1994.

Vol. 1594: M. Green, J. Murre, C. Voisin, Algebraic Cycles and Hodge Theory. Torino, 1993. Editors: A. Albano, F. Bardelli. VII, 275 pages. 1994.

Vol. 1595: R.D.M. Accola, Topics in the Theory of Riemann Surfaces. IX, 105 pages. 1994.

Vol. 1596: L. Heindorf, L. B. Shapiro, Nearly Projective Boolean Algebras. X, 202 pages. 1994.

Vol. 1597: B. Herzog, Kodaira-Spencer Maps in Local Algebra. XVII, 176 pages. 1994.

Vol. 1598: J. Berndt, F. Tricerri, L. Vanhecke, Generalized Heisenberg Groups and Damek-Ricci Harmonic Spaces. VIII, 125 pages. 1995.

Vol. 1599: K. Johannson, Topology and Combinatorics of 3-Manifolds. XVIII, 446 pages. 1995.

Vol. 1600: W. Narkiewicz, Polynomial Mappings. VII, 130 pages. 1995.

Vol. 1601: A. Pott, Finite Geometry and Character Theory. VII, 181 pages. 1995.

Vol. 1602: J. Winkelmann, The Classification of Three-dimensional Homogeneous Complex Manifolds. XI, 230 pages. 1995.

Vol. 1603: V. Ene, Real Functions – Current Topics. XIII, 310 pages. 1995.

Vol. 1604: A. Huber, Mixed Motives and their Realization in Derived Categories. XV, 207 pages. 1995.

Vol. 1605: L. B. Wahlbin, Superconvergence in Galerkin Finite Element Methods. XI, 166 pages. 1995.

Vol. 1606: P.-D. Liu, M. Qian, Smooth Ergodic Theory of Random Dynamical Systems. XI, 221 pages. 1995.

Vol. 1607: G. Schwarz, Hodge Decomposition – A Method for Solving Boundary Value Problems. VII, 155 pages. 1995.

Vol. 1608: P. Biane, R. Durrett, Lectures on Probability Theory. Editor: P. Bernard. VII, 210 pages. 1995.

Vol. 1609: L. Arnold, C. Jones, K. Mischaikow, G. Raugel, Dynamical Systems. Montecatini Terme, 1994. Editor: R. Johnson. VIII. 329 pages. 1995.

Vol. 610: A. S. Üstünel. An Introduction to Analysis on Wiener Space. X, 95 pages. 1995.

Vol. 1611: N. Knarr, Translation Planes. VI, 112 pages. 1995.

Vol. 1612: W. Kühnel, Tight Polyhedral Submanifolds and Tight Triangulations. VII, 122 pages. 1995.

Vol. 1613: J. Azéma, M. Emery, P. A. Meyer, M. Yor (Eds.). Séminaire de Probabilités XXIX. VI, 326 pages. 1995.

Vol. 1614: A. Koshelev, Regularity Problem for Quasilinear Elliptic and Parabolic Systems. XXI. 255 pages. 1995.

Vol. 1615: D. B. Massey. Lê Cycles and Hypersurface Singularities. XI, 131 pages. 1995.

Vol. 1616: I. Moerdijk, Classifying Spaces and Classifying Topoi. VII, 94 pages. 1995.

Vol. 1617: V. Yurinsky, Sums and Gaussian Vectors. XI, 305 pages. 1995.

Vol. 1618: G. Pisier, Similarity Problems and Completely Bounded Maps. VII, 156 pages. 1996.

Vol. 1619: E. Landvogt, A Compactification of the Bruhat-Tits Building. VII, 152 pages. 1996.

Vol. 1620: R. Donagi, B. Dubrovin, E. Frenkel, E. Previato, Integrable Systems and Quantum Groups. Montecatini Terme, 1993. Editors:M. Francaviglia, S. Greco. VIII. 488 pages 1996.

Vol. 1621: H. Bass, M. V. Otero-Espinar, D. N. Rockmore, C. P. L. Tresser. Cyclic Renormalization and Auto-morphism Groups of Rooted Trees. XXI. 136 pages. 1996.

Vol. 1622: E. D. Farjoun. Cellular Spaces, Null Spaces and Homotopy Localization. XIV, 199 pages. 1996.

Vol. 1623: H.P. Yap, Total Colourings of Graphs. VIII, 131 pages. 1996.

Vol. 1624: V. Brînzănescu, Holomorphic Vector Bundles over Compact Complex Surfaces. X, 170 pages. 1996.

Vol.1625: S. Lang, Topics in Cohomology of Groups. VII, 226 pages. 1996.

Vol. 1626: J. Azéma, M. Emery, M. Yor (Eds.), Séminaire de Probabilités XXX. VIII, 382 pages. 1996.

Vol. 1627: C. Graham, Th. G. Kurtz, S. Méléard, Ph. E. Protter, M. Pulvirenti, D. Talay, Probabilistic Models for Nonlinear Partial Differential Equations. Montecatini Terme, 1995. Editors: D. Talay, L. Tubaro. X, 301 pages. 1996.

Vol. 1628: P.-H. Zieschang, An Algebraic Approach to Association Schemes. XII, 189 pages. 1996.

Vol. 1629: J. D. Moore, Lectures on Seiberg-Witten Invariants. VII, 105 pages. 1996.

Vol. 1630: D. Neuenschwander, Probabilities on the Heisenberg Group: Limit Theorems and Brownian Motion. VIII, 139 pages. 1996.

Vol. 1631: K. Nishioka, Mahler Functions and Transcendence.VIII, 185 pages.1996.

Vol. 1632: A. Kushkuley, Z. Balanov, Geometric Methods in Degree Theory for Equivariant Maps. VII, 136 pages. 1996.

Vol.1633: H. Aikawa, M. Essén, Potential Theory – Selected Topics. IX, 200 pages.1996.

Vol. 1634: J. Xu, Flat Covers of Modules. IX, 161 pages. 1996.

Vol. 1635: E. Hebey, Sobolev Spaces on Riemannian Manifolds. X. 116 pages. 1996.

Vol. 1636: M. A. Marshall. Spaces of Orderings and Abstract Real Spectra. VI. 190 pages. 1996.

Vol. 1637: B. Hunt. The Geometry of some special Arithmetic Quotients. XIII, 332 pages. 1996.

Vol. 1638: P. Vanhaecke. Integrable Systems in the realm of Algebraic Geometry. VIII, 218 pages. 1996.

Vol. 1639: K. Dekimpe, Almost-Bieberbach Groups: Affine and Polynomial Structures. X, 259 pages. 1996.

Vol. 1640: G. Boillat, C. M. Dafermos, P. D. Lax, T. P. Liu, Recent Mathematical Methods in Nonlinear Wave Propagation. Montecatini Terme, 1994. Editor: T. Ruggeri. VII. 142 pages. 1996.

Vol. 1641: P. Abramenko, Twin Buildings and Applications to S-Arithmetic Groups. IX, 123 pages. 1996.

Vol. 1642: M. Puschnigg, Asymptotic Cyclic Cohomology. XXII, 138 pages. 1996.

Vol. 1643: J. Richter-Gebert, Realization Spaces of Polytopes. XI, 187 pages. 1996.

Vol. 1644: A. Adler, S. Ramanan, Moduli of Abelian Varieties. VI, 196 pages. 1996.

Vol. 1645: H. W. Broer, G. B. Huitema, M. B. Sevryuk, Quasi-Periodic Motions in Families of Dynamical Systems. XI. 195 pages. 1996.

Vol. 1646: J.-P. Demailly, T. Peternell, G. Tian, A. N. Tyurin, Transcendental Methods in Algebraic Geometry. Cetraro, 1994. Editors: F. Catanese, C. Ciliberto. VII. 257 pages. 1996.

Vol. 1647: D. Dias, P. Le Barz. Configuration Spaces over Hilbert Schemes and Applications. VII. 143 pages. 1996.

Vol. 1648: R. Dobrushin, P. Groeneboom, M. Ledoux, Lectures on Probability Theory and Statistics. Editor: P. Bernard. VIII. 300 pages. 1996.

Vol. 1649: S. Kumar, G. Laumon, U. Stuhler, Vector Bundles on Curves – New Directions. Cetraro, 1995. Editor: M. S. Narasimhan. VII, 193 pages. 1997.

Vol. 1650: J. Wildeshaus, Realizations of Polylogarithms. XI. 343 pages. 1997.

Vol. 1651: M. Drmota, R. F. Tichy, Sequences, Discrepancies and Applications. XIII, 503 pages. 1997.

Vol. 1652: S. Todorcevic, Topics in Topology. VIII, 153 pages. 1997.

Vol. 1653: R. Benedetti, C. Petronio, Branched Standard Spines of 3-manifolds. VIII, 132 pages. 1997.